Topics in Model Theory

Topics in Model Theory

Anand Pillay
University of Notre Dame, USA

World Scientific

NEW JERSEY • LONDON • SINGAPORE • BEIJING • SHANGHAI • HONG KONG • TAIPEI • CHENNAI • TOKYO

Published by

World Scientific Publishing Co. Pte. Ltd.
5 Toh Tuck Link, Singapore 596224
USA office: 27 Warren Street, Suite 401-402, Hackensack, NJ 07601
UK office: 57 Shelton Street, Covent Garden, London WC2H 9HE

Library of Congress Cataloging-in-Publication Data
Names: Pillay, Anand, author.
Title: Topics in model theory / Anand Pillay, University of Notre Dame, USA.
Description: New Jersey : World Scientific, [2024] | Includes bibliographical references and index.
Identifiers: LCCN 2024005299 | ISBN 9789811243806 (hardcover) |
ISBN 9789811243998 (paperback) | ISBN 9789811243813 (ebook for institutions) |
ISBN 9789811243820 (ebook for individuals)
Subjects: LCSH: Model theory. | Stability.
Classification: LCC QA9.7 .P553 2024 | DDC 511.3/4--dc23/eng/20240324
LC record available at https://lccn.loc.gov/2024005299

British Library Cataloguing-in-Publication Data
A catalogue record for this book is available from the British Library.

For any available supplementary material, please visit
https://www.worldscientific.com/worldscibooks/10.1142/12455#t=suppl

Printed in Singapore

This book is dedicated to the memory of my mother Patricia Pillay, who died on February 10th 2021, shortly after her 96th birthday.

Preface

This book *Topics in Model Theory* is based on two graduate courses I taught at the University of Notre Dame, in the Autumn of 2018, and in the Spring of 2021.

The general area is of course model theory. The 2018 course was on stability theory and the 2021 course on continuous logic (or continuous model theory). Each of the chapters in the book has its own fairly detailed introduction as well as references. But the material in the two courses is inter-related and in this preface I will give an overview.

Stability theory, the study of stable first order theories, was developed by Saharon Shelah, in connection with his successful program to classify first order theories according to whether or not their models can be described by (generalized) cardinal invariants. The subject and its machinery represented a new level of sophistication in model theory, although for a long time many people, including myself, believed that the technical machinery was only meaningful for the very small class of stable theories. Many developments since the mid 1990's have shown that this machinery, especially "forking" and its relatives, can make sense in and can be used to understand definability in broader and broader classes of first order theories, far beyond the stable ones. Also this understanding of definability in specific first order theories of interest has led to major applications of model theory to other parts of mathematics, Hrushovski being a key player. Within stability proper, the so-called "local theory" plays an important role. Here one only assumes stability (i.e. absence of the order property) of a single formula $\phi(x, y)$ and studies the Boolean algebra (of "ϕ-formulas") generated by instances $\phi(x, b)$ of ϕ (or if one wants, sets defined by such instances in a structure M), as well as ultrafilters on such a Boolean algebra, complete ϕ-types. An important aspect of the stability theory chapter is a modern account of this “local theory”. Keisler measures, generalizing complete types, are becoming more pervasive and important in model theory, especially since their role in solving my conjectures relating groups in o-minimal structures to compact Lie groups. The chapter finishes with an observation about Keisler measures on the Boolean algebra of ϕ-formulas for $\phi(x, y)$ stable, and how this can be essentially viewed as a restatement of a stable version of Szemeredi regularity.

Continuous logic (or continuous model theory) is about the adaptation of first order logic to a context where formulas are real-valued. A large part of the continuous logic chapter is devoted to an exposition of a formalism which has become fairly popular (partly so that I and the students in the course could learn it). But I also discuss other formalisms, as well as the extent to which continuous logic has origins and already exists within classical first order logic. The last part of the chapter studies local stability in continuous logic, which of course generalizes the material on local stability in the first chapter. I point out (following earlier observations of Ben Yaacov) the connection between local continuous stability and functional analysis, specifically Grothendieck's thesis. I derive the “fundamental theorem” of local continuous stability from a theorem of Grothendieck. The connection was already pointed out in the first chapter. It is amazing that the notion of stability appeared independently and much earlier in functional analysis, as well as in topological dynamics (weakly almost periodic dynamical systems). This is an example of the “unity of mathematics”, where the pervasiveness or depth of a notion is witnessed by its appearing independently in different areas. We point out how the stability of the inner product formula in Hilbert spaces has consequences for Keisler measures in classical first order theories. Again making the connection with the first chapter, the second chapter concludes with a discussion of Keisler measures and local stability in the continuous logic framework together with an application (without proof) to an analytic (or functional version) of stable regularity.

In any case this book represents in a sense my current view of aspects of these two topics, Stability Theory and Continuous Logic.

Acknowledgements

Through the roughly 5 year period from giving the 2018 course to preparing the text for publication, I received support from the National Science Foundation, grants DMS-1665035, DMS-1760212 and DMS-205427, whom

I thank. I would like to thank the Simons Foundation and Fields Institute, Toronto, for a Simons Distinguished Visitorship in the second half of 2021. Thanks to Imperial College, London, for a Nelder Fellowship in Summer 2022. Thanks again to the Simons Institute for their support (as Simons senior leader) during the Simons Semester STRUCTURES in Poland, Autumn 2023. Finally thanks to Elzbieta Pillay for her support throughout.

Anand Pillay
South Bend, January 9, 2024

Contents

Chapter 1

Stability Theory

1.1 Introduction

These are notes from the course "Topics in Stability Theory" given by me at Notre Dame in the autumn of 2018. They were originally written up by the graduate students enrolled in the course: Rachael Alvir, Nicolas Chavarria, Greg Cousins, Yayi Fu, Kyle Gannon, Grant Goodman, Leo Jimenez, Liling Ko, Justin Miller, and Jinhe (Vincent) Ye. Many thanks to them. Some of the material in the notes was used by me in a course on stability at the Fields Institute in Fall 2021, when I made a few mathematical corrections and additions, and benefitted from comments by the audience. In any case, I tinkered with the notes quite a bit, although they are probably still rather uneven. In particular there may still be some minor variations in notation, depending on who wrote up the notes in question.

Thanks also to Paul Larson for comments on and corrections to the original 2018 notes.

We will assume familiarity with basic model theory, for which the first parts of [TZ12] are good references.

The course first covered, in the preliminaries section, some tools required to develop stability theory, such as indiscernibles and imaginaries. Definability and generalizations were discussed, including formulas with values in a compact space (continuous logic). Then local stability was developed, and consequences for stable theories were then discussed.

The course then became more expository, starting with a survey of classification theory, and different strengthenings of stability. Then there is a survey of geometric stability theory, including Mordell–Lang. Finally, after a brief introduction to Keisler measures, the stable regularity lemma is proved, using previous tools.

The treatment of local stability is similar to that in the author's book [Pil96], but somewhat more "modern" from the author's point of view: the fundamental theorem relating definability and finite satisfiability, is done in the context of stability over a model, and the connection with Grothendieck's thesis is mentioned. The word forking is used as in Shelah's original definition. Of course the general theory is due to Shelah, see [She90].

Our referencing is rather uneven and we may refer to other works or books for more detailed and accurate references. Apologies to anybody concerned.

1.2 Preliminaries

We will use standard notation, such as in [Pil02] (which will eventually appear as part of an AMS book). We will start these notes by recalling some of these conventions.

The letter T will always denote a complete 1-sorted first order theory, in a language L. Models of the theory will be denoted by M, $N \models T$, and their subsets will be denoted by $A, B, \cdots$. Finite tuples of elements of models are denoted $\bar{a}, \bar{b}, \cdots$ (or sometimes, just $a, b, \ldots$ if there is no room for ambiguity). If $A \subseteq M$, then by $\mathrm{Th}(M, A)$ we mean the complete theory of M with a constant symbol for each element of A. (This language is called L_A.)

As is the usual in model theory, we will fix a large cardinal $\bar{\kappa}$, and a $\bar{\kappa}$-saturated and strongly $\bar{\kappa}$-homogeneous model $\overline{M}$ of T. We will sometimes refer to it as the monster model. All sets of parameters considered will usually be of size strictly smaller than $\bar{\kappa}$.

Note. With the above notation, any model of cardinality $\leqslant \bar{\kappa}$ will be isomorphic to an elementary substructure of $\bar{M}$. By convention, a *model* M denotes an elementary substructure of $\bar{M}$.

If $\Sigma(\bar{y})$ denotes a set of L_A-formulas, and $\varphi(\bar{y})$ denotes an L_A-formula, then $\Sigma(\bar{y}) \models \varphi(\bar{y})$ means that for all $\bar{b}$ in $\bar{M}$, if $\bar{M} \models \Sigma(\bar{b})$, then $\bar{M} \models \varphi(\bar{b})$.

Remark 1.2.1. If M is $\bar{\kappa}$-saturated and $\Sigma(\bar{y}) \models \varphi(\bar{y})$, then there exists a finite set of formula $\Sigma'(\bar{y}) \subset \Sigma(\bar{y})$ such that $\Sigma'(\bar{y}) \models \varphi(\bar{y})$.

Proof. Otherwise every finite subset of $\Sigma(\bar{y}) \cup \{\neg\varphi(\bar{y})\}$ will be consistent. By compactness, this set of formulas would then be consistent. Saturation of M yields that the set of formulas is realized in M, contradicting $\Sigma(\bar{y}) \models \varphi(\bar{y})$. $\square$

A subset X of $\bar{M}^n$ is said to be A-definable in $\bar{M}$ if there is $\varphi(\bar{x}) \in L_A$ such that $X = \{\bar{b} \in \bar{M} : \bar{M} \models \varphi(\bar{b})\}$. Finally, if $A \subseteq M$, and $\bar{b} \in M$, then the type of $\bar{b}$ over A in M is the set of formulas $\varphi(\bar{x}) \in L_A$ such that $M \models \varphi(\bar{b})$, and is denoted $\mathrm{tp}_M(\bar{b}/A)$. Note that this is the same as $tp_{\bar{M}}(\bar{b}/A)$, which we often just write as $\mathrm{tp}(\bar{b}/A)$.

1.2.1 *Indiscernibles*

Definition 1.2.2 (Indiscernibles). Let $(I, <)$ be a totally ordered set and $(\bar{a}_i : i \in I)$ a sequence of finite tuples from $\bar{M}$ of the same length. We say $(\bar{a}_i : i \in I)$ is *indiscernible* if for each $i_1 < \ldots < i_n$ and $j_1 < \ldots < j_n \in I$,

$$\mathrm{tp}_{\bar{M}}(\bar{a}_{i_1}, \ldots, \bar{a}_{i_n}) = \mathrm{tp}_{\bar{M}}(\bar{a}_{j_1}, \ldots, \bar{a}_{j_n}).$$

Equivalently, for all $\varphi(\bar{x}_1, \ldots, \bar{x}_n) \in L$,

$$\bar{M} \models \varphi(\bar{a}_{i_1}, \ldots, \bar{a}_{i_n}) \quad \text{iff} \quad \bar{M} \models \varphi(\bar{a}_{j_1}, \ldots, \bar{a}_{j_n}).$$

Note. The above definition can be extended to make sense for any model M and also for formulas over $A \subseteq M$.

Example. Let $M = (\mathbb{Q}, <)$, and $I = (\mathbb{Q}, <)$. Then I is indiscernible by quantifier elimination for dense linear orders.

Theorem 1.2.3. *Compactness lets us "stretch" indiscernibles. Formally, let $(a_i : i \in \omega)$ be indiscernible in $\bar{M}$, and $(I, <)$ an ordering of cardinality smaller than $\bar{\kappa}$. Then there exists an indiscernible $(b_i : i \in I)$ in $\bar{M}$ such that $\forall i_1 < \ldots < i_n \in I$,*

$$\mathrm{tp}_{\bar{M}}(a_1, \ldots, a_n) = \mathrm{tp}_{\bar{M}}(b_{i_1}, \ldots, b_{i_n}).$$

Proof. Introduce new constant symbols $c_i, i \in I$ and let

$$\Sigma = \{\varphi(c_1, \ldots, c_n) : \bar{M} \models \varphi(a_1, \ldots, a_n)\},$$

$$\Sigma' = \{\varphi(c_1, \ldots, c_n) \leftrightarrow \varphi(c_{i_1}, \ldots, c_{i_n}) : i_1 < \ldots < i_n \in I\}.$$

Then $\Sigma \cup \Sigma' \cup T$ is consistent by compactness, i.e. has a model. Take such a model, of cardinality $\leqslant \bar{\kappa}$. By $\bar{\kappa}$-saturation and strong $\bar{\kappa}$-homogeneity, this model is isomorphic to an elementary substructure of $\bar{M}$. Then

$$\bar{M} \models \varphi(a_1, \ldots, a_n) \leftrightarrow \bar{M} \models \varphi(c_1, \ldots, c_n) \qquad \text{(by } \Sigma'\text{)}$$

$$\leftrightarrow \bar{M} \models \varphi(c_{i_1}, \ldots, c_{i_n}). \qquad \text{(by } \Sigma\text{)}$$

$\square$

The same holds for indiscernible sequences of (even infinite) tuples, in place of elements.

Indiscernible sequences are a fundamental tool of model theory, and there are many ways to obtain them. In what follows, we will discuss three such methods: Ramsey's theorem, coheirs, and the Erdös–Rado theorem.

Fact (Ramsey, extended). *Let $n_1, \ldots, n_r < \omega$. For each $i = 1, \ldots, r$, let $X_{i,1}$, $X_{i,2}$ be a partition of $\omega^{[n_i]}$, the set of n_i-element subsets of $\mathbb{N}$. Then there is an infinite subset $Y \subseteq \omega$ which is homogeneous, i.e. for all $i = 1, \ldots, r$, either $Y^{[n_i]} \subseteq X_{i,1}$ or $Y^{[n_i]} \subseteq X_{i,2}$.*

We can apply Ramsey's theorem to obtain indiscernible sequences.

Definition 1.2.4. Let $A \subseteq \bar{M}$ be small and $\Sigma(\bar{x})$ be a collection of L_A-formulas in the variables $\bar{x}$. We say "$\Sigma(\bar{x})$ is consistent" if $\Sigma(\bar{x}) \cup \mathrm{Th}(\bar{M}, A)$ is consistent.

Fact 1.2.5. *$\Sigma(\bar{x})$ is consistent iff every finite subset $\Sigma'(\bar{x}) \subseteq \Sigma(\bar{x})$ is realized in $\bar{M}$.*

Proof. If $\Sigma(\bar{x})$ is consistent, then by saturation it is realized in $\bar{M}$, and in particular every finite subset $\Sigma'(\bar{x}) \subseteq \Sigma(\bar{x})$ will also be realized in $\bar{M}$. For the converse, if every finite subset is realized, every finite subset is consistent. Then from compactness Σ is consistent. □

Proposition 1.2.6. *For each $n \in \omega$, let $\Sigma_n(x_1, \ldots, x_n)$ be a collection of L-formulas in variables $x_1, \ldots, x_n$. Suppose that there are $a_1, a_2, \cdots \in \bar{M}$ such that*

$$\bar{M} \models \Sigma_n(a_{i_1}, \ldots, a_{i_n}), \quad \forall i_1 < \ldots < i_n < \omega.$$

Then there exists an indiscernible $(b_i : i \in \omega)$ in $\bar{M}$ such that

$$\bar{M} \models \Sigma_n(b_{i_1}, \ldots, b_{i_n}), \quad \forall i_1 < \ldots < i_n < \omega.$$

Example. Suppose $\Sigma_2 = \{x_1 \neq x_2\}$. Then the previous proposition yields the existence of infinite indiscernible sequences.

Proof of Proposition 1.2.6. Consider the following set of L-formulas

$$\begin{aligned}\Gamma(x_1, x_2, \ldots) = \{&\varphi(x_{i_1}, \ldots, x_{i_n}) \leftrightarrow \varphi(x_{j_1}, \ldots, x_{j_n}) : \\ &i_1 < \ldots < i_n,\ j_1 < \ldots < j_n \in \omega, \varphi \in L\} \\ &\cup \bigcup_n \Sigma_n(x_1, \ldots, x_n).\end{aligned}$$

By Fact 1.2.5, it is enough to prove that every finite subset of Γ is consistent. Let Γ' be a finite subset of Γ. By choosing n large enough, we can assume the only variables appearing in Γ' are $x_1, \ldots, x_n$. Let $\varphi_1, \ldots, \varphi_r$ be the L-formulas appearing in Γ'. For $i = 1, \ldots, r$, let

$$X_{i,1} = \{(j_1, \ldots, j_n) : \quad j_1 < \ldots < j_n \in \omega, \quad \bar{M} \models \varphi_i(a_{j_1}, \ldots, a_{j_n})\},$$

$$X_{i,2} = \{(j_1, \ldots, j_n) : \quad j_1 < \ldots < j_n \in \omega, \quad \bar{M} \models \neg\varphi_i(a_{j_1}, \ldots, a_{j_n})\}.$$

By Ramsey's theorem, there exists an infinite $Y \subseteq \mathbb{N}$ such that for all $i = 1, \ldots, r$, $Y^{[n_i]}$ is either contained in $X_{i,1}$ or in $X_{i,2}$. Write $Y = \{k_1 < k_2 < \ldots\}$. Interpret each x_i as a_{k_i} to satisfy Γ'. □

Note. Proposition 1.2.6 also works when we consider tuples instead of elements, and when we consider formulas over some small fixed $A \subset \bar{M}$.

Another way to obtain indiscernibles is via the use of special types, called coheirs. We will define them now. First recall that if A is any set of parameters, then $S_{\bar{x}}(A)$ denotes the class of complete types over A.

Definition 1.2.7. Let $M \prec N \prec \bar{M}$ be models, and $p(\bar{x}) \in S_{\bar{x}}(N)$. We say p is finitely satisfiable in M, or $p(\bar{x})$ is a *coheir* of $p{\restriction}M \in S_{\bar{x}}(M)$, if every $\varphi(\bar{x}) \in p(\bar{x})$ is satisfied by some $\bar{a} \in M$, or equivalently, every finite subset of $p(\bar{x})$ is realized by some $\bar{a} \in M$.

Note. Any $\varphi(\bar{x}) \in p$ is realized in N as $\varphi(\bar{x}) \in L_N$ and $N \prec \bar{M}$ and $\varphi(\bar{x})$ is realized in $\bar{M}$, but there is no reason to expect φ to be realized in M.

Remark. $p(\bar{x}) \in S_{\bar{x}}(N)$ is finitely satisfiable (f.s.) in M if and only if $p(\bar{x})$ is in the topological closure of $\{\mathrm{tp}(\bar{a}/N) : \bar{a} \in M\}) \subseteq S_{\bar{x}}(N)$.

Proof. Suppose first that $p(\bar{x})$ is finitely satisfiable. Let $\varphi(\bar{x}) \in p(\bar{x})$, and let $[\varphi(\bar{x})]$ be the corresponding open set. Then φ is realized in M, say by $\bar{a} \in M$. Hence $\mathrm{tp}(\bar{a}/N) \in [\varphi(\bar{x})]$.

For the other direction, assume $p(\bar{x})$ is in the closure of $\{\mathrm{tp}(\bar{a}/N) : \bar{a} \in M\}) \subseteq S_{\bar{x}}(N)$. Let $\varphi(\bar{x}) \in p(\bar{x})$, then by assumption, there is $\bar{a} \in M$ such that $\mathrm{tp}(\bar{a}/N) \in \varphi(\bar{x})$, so we obtain $M \models \varphi(\bar{a})$. □

Lemma 1.2.8. *Suppose $p(\bar{x}) \in S_{\bar{x}}(M)$ and $M \prec N$. Then there is $p'(\bar{x}) \in S_{\bar{x}}(N)$ such that $p \subseteq p'$ and p' is finitely satisfiable in M.*

Proof. Consider $\Gamma(\bar{x}) = p(\bar{x}) \cup \{\neg\varphi(\bar{x}) : \varphi(\bar{x}) \in L_N$ and not realized in $M\}$. It is enough to show that $\Gamma(\bar{x})$ is consistent because any extension of such Γ to a complete type $p'(\bar{x}) \in S_{\bar{x}}(N)$ will work. Let Γ' be a finite subset of Γ, written in the form $\Gamma' = \{\psi(\bar{x}), \neg\varphi_1(\bar{x}), \ldots, \neg\varphi_r(\bar{x})\} \in p$. Then any solution $\bar{a}$ of ψ in M satisfies Γ', as the $\varphi_1, \cdots, \varphi_r$ are not realized in M. $\square$

Remark. The following is a proof of the above fact using analysis. Let i_M denote the map from $M^{\bar{x}}$ to $S_{\bar{x}}(M)$ such that $m \mapsto \mathrm{tp}(m/M)$. We define $i_N : M^{\bar{x}} \to S_{\bar{x}}(N)$ similarly. Let r denote the restriction map from $S_{\bar{x}}(N) \to S_{\bar{x}}(M)$. Note that $r \circ i_N = i_M$, and the set of types in $S_{\bar{x}}(N)$ that are finitely satisfiable in M is exactly the closure of $i_N(M^{\bar{x}})$ in $S_{\bar{x}}(N)$. Hence its image under the restriction map is closed. However the image must contain $i_M(M^{\bar{x}})$, which is dense in $S_{\bar{x}}(M)$. Therefore it must be onto, which proves the desired result.

Note. Lemma 1.2.8 also works when we consider formulas over some $A \subset M$, i.e. if $A \subset M$ and $p(\bar{x}) \in S_{\bar{x}}(M)$ is finitely satisfiable in A, then for any model $N \succ M$, there is $p'(\bar{x}) \in S_{\bar{x}}(N)$ such that $p \subseteq p'$ and p' is finitely satisfiable in A.

Proposition 1.2.9. *Let $p(\bar{x}) \in S_{\bar{x}}(M)$, $N \succ M$ be $|M|^+$-saturated, and $p'(\bar{x}) \in S_{\bar{x}}(N)$ a coheir of p. Let $\bar{a}_1, \bar{a}_2, \ldots \in N$ be defined as follows:*

$\bar{a}_1$ *realizes* $p(\bar{x})$,
$\bar{a}_2$ *realizes* $p'(\bar{x})\restriction(M, \bar{a}_1)$,
$\bar{a}_3$ *realizes* $p'(\bar{x})\restriction(M, \bar{a}_1, \bar{a}_2)$,
$\ldots$.

Then $(\bar{a}_i : i \in \omega)$ is indiscernible over M.

Proof. We prove by induction on k that for any $n \leqslant k$ and $i_1 < \ldots < i_n \leqslant k$ and $j_1 < \ldots < j_n \leqslant k$, we have

$$\mathrm{tp}_M(\bar{a}_{i_1}, \ldots, \bar{a}_{i_n}/M) = \mathrm{tp}_M(\bar{a}_{j_1}, \ldots, \bar{a}_{j_n}/M).$$

Assume this is true for k, and consider $k+1$. Let $i_i < \ldots < i_n \leqslant k$, $j_1 < \ldots < j_n \leqslant k$. We need to show that

$$\mathrm{tp}_M(\bar{a}_{i_1}, \ldots, \bar{a}_{i_n}, \bar{a}_{k+1}/M) = \mathrm{tp}_M(\bar{a}_{j_1}, \ldots, \bar{a}_{j_n}, \bar{a}_{k+1}/M).$$

Consider a formula $\varphi(\bar{x}_1, \ldots, \bar{x}_n, \bar{x}_{n+1}) \in L_M$. Assume for a contradiction that

$$M \models \varphi(\bar{a}_{i_1}, \ldots, \bar{a}_{i_n}, \bar{a}_{k+1}) \wedge \neg\varphi(\bar{a}_{j_1}, \ldots, \bar{a}_{j_n}, \bar{a}_{k+1}).$$

But $\mathrm{tp}(\bar{a}_{k+1}/M, \bar{a}_1, \ldots, \bar{a}_k)$ is finitely satisfiable in M, so there is $\bar{a}' \in M$ such that

$$M \models \varphi(\bar{a}_{i_1}, \ldots, \bar{a}_{i_n}, \bar{a}') \wedge \neg\varphi(\bar{a}_{j_1}, \ldots, \bar{a}_{j_n}, \bar{a}'),$$

which contradicts the induction hypothesis.

(Note that this still works of $p(\bar{x})$ is realized in M, in which case the $\bar{a}_i$ are all the same.) $\square$

Finally, the Erdös–Rado theorem allows us to obtain indiscernibles. We will not expand on this, but let us state the following:

Proposition 1.2.10. *Given T, there exists λ (assumed $< \bar{\kappa}$) such that if $(\bar{a}_i : i < \lambda)$ is a sequence of finite tuples in $\bar{M}$, then there exists an indiscernible sequence $(\bar{b}_i : i < \omega)$ in $\bar{M}$ (tuples of the same length as the $\bar{a}_i$) such that for all $n < \omega$, there are $\alpha_0 < \ldots < \alpha_n < \lambda$ satisfying*

$$\mathrm{tp}_{\bar{M}}(\bar{b}_0, \ldots, \bar{b}_n) = \mathrm{tp}_{\bar{M}}(\bar{a}_{\alpha_0}, \ldots, \bar{a}_{\alpha_n}).$$

1.2.2 *Definability and generalizations*

Let $A \subseteq \bar{M}$ be small. Recall an A-definable set in $\bar{M}$ is a subset $X \subseteq \bar{M}^n$ defined by an L_A-formula $\varphi(x_1, \ldots, x_n)$. We will call $\varphi(\bar{x}), \psi(\bar{x}) \in L_{\bar{M}}$ equivalent if $\bar{M} \models (\forall \bar{x})\,(\varphi(\bar{x}) \leftrightarrow \psi(\bar{x}))$. That is, the formulas φ, ψ define the same set.

Lemma 1.2.11. *Let $X \subseteq \bar{M}^n$ be definable (with parameters in $\bar{M}$). Then X is definable over A (i.e. X is definable by an L_A-formula) iff X is* $\mathrm{Aut}(\bar{M}/A)$*-invariant (where $\mathrm{Aut}(\bar{M}/A)$ denotes the group of automorphisms of $\bar{M}$ which fix A pointwise).*

Proof. The left to right direction is immediate: If X is defined by $\varphi(\bar{x}, \bar{a})$ with $\bar{a}$ from A, and σ is an automorphism of $\bar{M}$, then $\sigma(X)$ is defined by $\varphi(\bar{x}, \sigma(\bar{a}))$ which coincides with $\varphi(\bar{x}, \bar{a})$ if σ fixes A pointwise.

To prove the converse, we need to use strong $\bar{\kappa}$-homogeneity of $\bar{M}$. Suppose X is defined by $\varphi(\bar{x}, \bar{b})$, where $\bar{b} \in M$ are parameters.

Claim. *Given $\bar{b}' \in \bar{M}$, if* $\mathrm{tp}(\bar{b}'/A) = \mathrm{tp}(\bar{b}/A)$, *then $\varphi(\bar{x}, \bar{b}')$ is equivalent to $\varphi(\bar{x}, \bar{b})$.*

Proof. By strong $\bar{\kappa}$-homogeneity of $\bar{M}$, there is some $\sigma \in \mathrm{Aut}(\bar{M}/A)$ such that $\sigma(\bar{b}) = \bar{b}'$. Then $\varphi(\bar{x}, \bar{b}')$ defines $\sigma(X)$. But $\sigma(X) = X$ by assumption. So $\varphi(\bar{x}, \bar{b}')$ is equivalent to $\varphi(\bar{x}, \bar{b})$. □

Let $p(\bar{y}) = \mathrm{tp}_{\bar{M}}(\bar{b}/A)$. The claim yields $p(\bar{y}) \models (\forall \bar{x})\,(\varphi(\bar{x}, \bar{y}) \leftrightarrow \varphi(\bar{x}, \bar{b}))$. Then compactness yields some $\psi(\bar{y}) \in p(\bar{y})$, $\psi \in L_A$ such that

$$\psi(\bar{y}) \models (\forall \bar{x})\,(\varphi(\bar{x}, \bar{y}) \leftrightarrow \varphi(\bar{x}, \bar{b})).$$

So X can be defined by the L_A-formula

$$\theta(\bar{x}) = (\exists \bar{y})\,(\varphi(\bar{x}, \bar{y}) \wedge \psi(\bar{y})) \in L_A.$$

□

The equivalence of definability and invariance under automorphisms can be generalized, as shown in the following definition and lemma.

Definition 1.2.12. $X \subseteq \bar{M}^n$ is definable almost over A if there is an A-definable equivalence relation E on $\bar{M}^n$ with finitely many classes and X is a union of some E-classes.

Lemma 1.2.13. *Let $X \subseteq \bar{M}^n$ be definable, and $A \subset \bar{M}$ be small. The following are equivalent:*

(i) X is definable almost over A.
(ii) $|\{\sigma(X) : \sigma \in \mathrm{Aut}(\bar{M}/A)\}| < \omega$.
(iii) $|\{\sigma(X) : \sigma \in \mathrm{Aut}(\bar{M}/A)\}| < \bar{\kappa}$.

Proof. (ii) $\Rightarrow$ (iii) is immediate.

(i) $\Rightarrow$ (ii) also follows from the definition: As E is an A-definable equivalence relation then it is invariant under automorphisms fixing X pointwise, so if X is a union of E-classes then so is $\sigma(X)$ for any $\sigma \in Aut(\bar{M}/A)$. If E has only finitely many classes then there are only finitely many possibilities for $\sigma(X)$, completing the proof.

(ii) $\Rightarrow$ (i): Let $\varphi(\bar{x}, \bar{b}) \in L_A$ define X and let $p(\bar{y}) = \mathrm{tp}(\bar{b}/A)$. Given $\sigma \in \mathrm{Aut}(\bar{M}/A)$, $\sigma(X)$ is defined by $\varphi(\bar{x}, \sigma(\bar{b}))$. Moreover for any realization $\bar{b}'$ of $p(\bar{y})$, there is, by the strong homogeneity assumption on $\bar{M}$, some $\sigma \in Aut(\bar{M}/A)$ such that $\bar{b}' = \sigma(\bar{b})$. Hence by our assumption (ii) there are realizations $\bar{b}_1, ..., \bar{b}_n$ of p such that

$$p(\bar{y}) \models (\forall \bar{x}) \bigvee_{i \leqslant n} [\varphi(\bar{x}, \bar{y}) \leftrightarrow \varphi(\bar{x}, \bar{b}_i)].$$

Then compactness yields some $\psi(\bar{y}) \in p(\bar{y})$ such that

$$\psi(\bar{y}) \models (\forall \bar{x}) \bigvee_{i \leqslant n} [\varphi(\bar{x}, \bar{y}) \leftrightarrow \varphi(\bar{x}, \bar{b}_i)]. \qquad (*)$$

Define the equivalence relation $E(\bar{x}_1, \bar{x}_2)$ by the formula

$$\theta(\bar{x}_1, \bar{x}_2) = (\forall \bar{y})[\psi(\bar{y}) \rightarrow (\varphi(\bar{x}_1, \bar{y}) \leftrightarrow \varphi(\bar{x}_2, \bar{y}))] \in L_A.$$

Then E has only finitely many equivalence classes from $(*)$, and X is a union of E-classes.

(iii) $\Rightarrow$ (ii): Assume for contradiction that for some infinite cardinal $\lambda < \bar{\kappa}$ we have:

$$|\{\sigma(X) : \sigma \in \mathrm{Aut}(\bar{M}/A)\}| = \lambda.$$

As before we assume X to be defined by $\varphi(\bar{x}, \bar{b})$ and $p(\bar{y}) = tp(\bar{b}/A)$.

Let $\bar{y}_\alpha$ for $\alpha < \bar{\kappa}$ be new variables, and consider the collection of formulas $\Sigma((\bar{y}_\alpha)_{\alpha<\bar{\kappa}})$ over A: $\cup_{\alpha<\bar{\kappa}} p(\bar{y}_\alpha) \cup \{\neg\forall\bar{x}(\varphi(\bar{x}, \bar{y}_\alpha) \leftrightarrow \varphi(\bar{x}, \bar{y}_\beta)) : \alpha < \beta < \bar{\kappa}\}$.

Then, as in the discussion in the proof of (ii) implies (i) above, this set of formulas is consistent, and by $\bar{\kappa}$-saturation of $\bar{M}$, Σ is realized in $\bar{M}$, yielding that

$$|\{\sigma(X) : \sigma \in \mathrm{Aut}(\bar{M}/A)\}| = \bar{\kappa}$$

contradicting (iii). □

Recall that $S_{\overline{x}}(A)$ is the set of complete types in free variables $\overline{x}$ with parameters from A. (Equivalently, the complete types in the language L_A in free variables $\overline{x}$ extending the L_A-theory $Th(\overline{M})$.) Since $\overline{M}$ is saturated and A is small, this is also the set of all $\mathrm{tp}_{\overline{M}}(\overline{a}/A)$ for some $\overline{a} \in \overline{M}$.

The topology on $S_{\overline{x}}(A)$ is generated by the basic open sets

$$[\varphi(\overline{x})] = \{p(\overline{x}) \in S_{\overline{x}}(A) : \varphi(\overline{x}) \in p(x)\}.$$

Since types are complete, these basic open sets are also closed because we have $S_{\bar{x}}(A)\backslash[\varphi(\overline{x})] = [\neg\varphi(\overline{x})]$. With this topology, $S_{\overline{x}}(A)$ is the Stone space of the Boolean algebra of L_A-formulas $\varphi(\overline{x})$ up to equivalence in $\overline{M}$. It is a compact, Hausdorff, totally disconnected space, also known as a profinite space (as it is an inverse limit of finite spaces).

Proposition 1.2.14.

1. *Formulas $\varphi(\overline{x})$, $\psi(\overline{x}) \in L_A$ are equivalent if and only if $[\varphi(\overline{x})] = [\psi(\overline{x})]$ in $S_{\overline{x}}(A)$.*
2. *The clopen subsets of $S_{\overline{X}}(A)$ are precisely the basic clopen sets.*
3. *Clopen subsets X of $S_{\overline{x}}(A)$ correspond exactly to continuous functions $f\colon S_{\overline{x}}(A) \to 2$, where $f(p(\overline{x})) = 1$ if $p(\overline{x}) \in X$ and 0 otherwise.*
4. *The A-definable subsets of $\overline{M}^n$ are in one-to-one correspondence with continuous functions from $S_{\overline{x}}(A)$ to 2.*

Proof.

1. Suppose $\varphi(\overline{x})$ and $\psi(\overline{x})$ are equivalent. Then $\overline{M} \models \forall\overline{x}$ $(\varphi(\overline{x}) \leftrightarrow \psi(\overline{x}))$. In particular, $[\psi(\overline{x})] \subseteq [\varphi(\overline{x})]$ and $\varphi(\overline{x}) \subseteq [\psi(\overline{x})]$. Indeed, suppose there was a type in one but not the other. Then by consistency, it is realized in the saturated model $\overline{M}$. That would imply that there is an $\overline{a} \in \overline{M}$ such that, for example, $\overline{M} \models \varphi(\overline{a}) \wedge \neg\psi(\overline{a})$, a contradiction.

Suppose $[\varphi(\overline{x})] = [\psi(\overline{x})]$, but they are not equivalent. Then $\overline{M} \models \exists\overline{x}\ \varphi(\overline{x}) \wedge \neg\psi(\overline{x})$. Let $\overline{a} \in \overline{M}$ be a witness. Then $\mathrm{tp}_{\overline{M}}(\overline{a}/A) \in [\varphi(\overline{x})]$ but not is not in $[\psi(\overline{x})]$, a contradiction. Thus they must be equivalent in $\overline{M}$.

2. We only need to show that any clopen set is a basic clopen set. Let $X \subseteq S_{\overline{x}}(A)$ be a clopen set. Hence X and its complement are closed, so each is the intersection of some collection of basic clopen sets represented by sets P_0 and P_1 of L_A-formulas. Then $P_0 \cup P_1$ is inconsistent because

$$\bigcap_{\varphi(\overline{x}) \in P_0 \cup P_1} [\varphi(\overline{x})] = X \cap X^c = \varnothing.$$

Therefore, by compactness there are finite subsets $\{\varphi_0(\overline{x}), \ldots, \varphi_k(\overline{x})\} \subseteq P_0$ and $\{\psi_0(\overline{x}), \ldots, \psi_j(\overline{x})\} \subseteq P_1$ whose union is inconsistent. Therefore the L_A-formula $\tau(\overline{x})$ defined via

$$\bigwedge_{i \in k+1} \varphi_i(\overline{x}) \wedge \bigwedge_{i \in j+1} \neg\psi_i(\overline{x})$$

has $[\tau(\overline{x})] = X$, so X is a basic clopen set as desired.

3. Suppose $f : S_{\overline{x}}(A) \to 2$ is as stated in the proposition. Then $f^{-1}(\{1\})$ is open because 2 has the discrete topology and f is continuous. Similarly, $f^{-1}(\{0\}) = \overline{f^{-1}(\{1\})}$ is open, so $f^{-1}(\{1\})$ is clopen. Note that $f^{-1}(\{1\})$ trivially gives rise to f as described in the proposition.
Now suppose X is a clopen set and define $f_X : S_{\overline{x}}(A) \to 2$ as described. Clearly f_X is continuous, as $f^{-1}(2) = S_{\overline{x}}(A)$, $f^{-1}(\varnothing) = \varnothing$, $f^{-1}(\{0\}) = X^c$, and $f^{-1}(\{1\}) = X$ are all open because X is clopen.

4. By the first part of the proposition, definable sets are in one-to-one correspondence with basic clopen subsets of $S_{\overline{x}}(A)$. By the second part, basic clopen sets are exactly all of the clopen subsets, so definable sets are in one-to-one correspondence with clopen sets. By the third part, clopen sets are in one-to-one correspondence with continuous functions $f : S_{\overline{x}}(A) \to 2$, so definable sets are in one-to-one correspondence with these continuous functions.

□

We can extend this idea to the general setting of continuous logic, which we shall introduce next. The above proposition shows that traditional logic is a special case of the following:

Definition 1.2.15.

1. Let C be a (Hausdorff) topological space. A C-valued formula in $\overline{x}$ over A is a continuous function $\varphi : S_{\overline{x}}(A) \to C$. Note that the image of φ is compact, so C may be assumed to be compact.
2. By a CL-formula (continuous logic) formula over A, we mean an $\mathbb{R}$-valued formula over A. Alternatively, the map can be into $[0, 1]$ or $\mathbb{R}_{\geqslant 0}$.

Remark 1.2.16.

- Suppose $\varphi : S_{\overline{x}}(A) \to C$ is a C-valued formula in $\overline{x}$ over A. Then φ gives rise to a map from $\overline{M}^n$ to C given by $f_\varphi(\overline{b}) = \varphi(\mathrm{tp}_{\overline{M}}(\overline{b}/A))$. We call f_φ an A-definable function from $\overline{M}^n$ to C.
- A definable function from $\overline{M}^n$ to C is an A-definable function from $\overline{M}^n$ to C for some small $A \subseteq \overline{M}$.
- For M an arbitrary model of T, a definable function from M^n to C is a function $f : M^n \to C$ which lifts to a continuous function from $S_{\overline{x}}(M)$ to C.

Recall that Lemma 1.2.11 showed that being a definable set and being invariant under automorphism are the same thing. This result adapts to continuous logic too.

Lemma 1.2.17. *Let f be a definable function from $\overline{M}^n$ to C. Let $A \subseteq \overline{M}$ be small. Then f is A-definable if and only if f is $Aut(\overline{M}/A)$-invariant, where the action of $Aut(\overline{M}/A)$ on f is $\sigma(f)(\overline{b}) = f(\sigma^{-1}(\overline{b}))$.*

Proof. Notice one direction is trivial. If f is A-definable, then f is $Aut(\overline{M}/A)$-invariant. We prove the other direction. Assume that f is $Aut(\overline{M}/A)$-invariant and definable. Then, it is B-definable for some small $B \subset \overline{M}$. Without loss of generality, we can assume that $A \subseteq B$. Let $f_B : S_{\overline{x}}(B) \to C$ be the map lifting f to $S_{\bar{x}}(B)$, that is, for all $p \in S_{\bar{x}}(B)$, we have $f_B(p) = f(a)$ for some (any) $a \models p$. Since f is B-definable, the map f_B is continuous.

Similarly, since f is A-invariant, the map $f_A : S_{\overline{x}}(A) \to C$ defined via $f_A(q) = f(c)$ where $c \models q$ is well-defined. Let $\pi_{B,A} : S_{\overline{x}}(B) \to S_{\overline{x}}(A)$ be the natural restriction map. Then, the following diagram commutes:

$$\begin{array}{ccc} S_{\overline{x}}(B) & & \\ \downarrow & \searrow & \\ S_{\overline{x}}(A) & \to & C \end{array}$$

The map $\pi_{B,A}$ is a topological quotient map, meaning that $\Omega \subset S_{\bar{x}}(B)$ is open if and only if $\pi_{B,A}^{-1}(\Omega)$ is open. Indeed, it is continuous, surjective and closed, as types spaces are compact Hausdorff.

By the universal property of quotient maps and continuity of f_B, we then obtain continuity of f_A. □

Example 1.2.18. Ordinary first order formulas and definable sets are examples, with $C = \{0, 1\}$.

Example 1.2.19. Consider the identity from $S_{\bar{x}}(A)$ to itself. Then this induces the tautological map that takes a tuple to its type, i.e. $f(\bar{b}) = \text{tp}_{\overline{M}}(\bar{b}/A)$.

Example 1.2.20. Let $T = RCOF = Th(\mathbb{R}, +, x, 0, 1, -, <)$. This theory admits quantifier elimination, that is all formulas are equivalent to some quantifier free formula defining a semi-algebraic set. Recall that a semi-algebraic set is defined by a finite disjunction of formulas of the form $f(\bar{x}) = 0 \wedge g(\bar{x}) > 0$.

A real closed field is a model of T, equivalently any ordered field with the intermediate value property. In any such model M, the absolute value function on M is definable over the empty set in the usual first order sense of definability.

In the context of real closed fields, we can also produce CL-formulas.

Example 1.2.21. Let $C = [-1, 1] \subseteq \mathbb{R}$ with the subspace topology, and let M be a model of $T = RCOF$ which extends (or contains) the standard model $\mathbb{R}$. Define the function $f : M \to C$ as follows:

If $|x|_M > 1$, then $f(x) = 0$. Otherwise, $f(x)$ is the unique element r of C such that for all $n \in \mathbb{N}\backslash\{0\}$ we have $|x - r|_M < \frac{1}{n}$.

We first claim that such an r exists and is unique. Assuming two such reals exist for a given x, equality follows immediately from the triangle inequality and the least upper bound property. Therefore we only need to show existence.

Suppose $|x|_M \leqslant 1$. Then since $M \models T$, $\mathbb{R}$ can be embedded in $\overline{M}$. Therefore, consider $\{r \in \mathbb{R} : r < x\}$. But $\mathbb{R}$ is complete as a metric space and $|x|_M \leqslant 1$, so this set has a supremum r in $\mathbb{R}$. Notice that r has the required property: if not, there is some $n \in \mathbb{N}\backslash\{0\}$ with $|x - r|_M \geqslant \frac{1}{n}$, and there are two cases:

- $r < x$: Then $|x - r|_M = x - r \geqslant \frac{1}{n}$, so $r + \frac{1}{2n} < x$, contradicting the fact that r is the supremum of the aforementioned set.

- $x < r$: Then $|x - r|_M = r - x \geqslant \frac{1}{n}$, so $r - \frac{1}{2n} > x$, contradicting the fact that r is the supremum.

We still need to show that this map $f(x)$, which we shall now denote as $st(x)$ for the "standard part of x," is definable in the sense mentioned above.

Note that any rational number q is definable over the empty set. Hence, for any type $p(x) \in S_1(M)$ with $|x| \leqslant 1$, we can consider the set $L_p = \{q \in \mathbb{Q}, q < x \in p\}$. It is a downward closed set, and it has a least upper bound, as it is bounded from above by 1.

This allows us to define the map:

$$\begin{aligned} f^* :& S_1(M) \to [0,1] \\ & p(x) \to \sup L_p \text{ if } |x| \leqslant 1 \in p(x) \\ & p(x) \to 0 \text{ else }. \end{aligned}$$

Then this map clearly induces $st(x)$ by the above argument since the rationals are dense in the reals. Furthermore, it is continuous: without loss of generality assume $0 < a < b < 1 \in C$ (the other cases are similar). Then

$$(f^*)^{-1}((a,b)) = \bigcup_{q<r\in\mathbb{Q}\cap(a,b)} [q < x \leqslant r].$$

Therefore the inverse image of a basic open set is open, so the function is continuous. Therefore $st(x)$ is definable as desired.

The interesting thing to note with this example is that the standard part map gives us a way to recover the usual topology on $\mathbb{R}$ from the Stone topology on type spaces.

1.2.3 *Imaginaries and T^{eq}*

Recall that $\mathrm{acl}_M(A)$ (the algebraic closure of A in M) is the set of $b \in M$ such that there exist an L_A-formula $\varphi(x)$ with $M \models \varphi(b)$ and $M \models \exists^{\leqslant k} x\ \varphi(x)$ for some $k > 0$. Similarly, the definable closure of A in M, denoted $\mathrm{dcl}_M(A)$, is the set of $b \in M$ such that there exist an L_A-formula $\varphi(x)$ with $M \models \varphi(b)$ and $M \models \exists^{\leqslant 1} x\ \varphi(x)$. The formal definition above also makes sense for b a finite tuple of elements. When we pass to T^{eq} and M^{eq} as will do below, the distinction between elements and finite tuples is blurred, although the distinction is recovered if we work inside a given "sort". Anyway in the one-sorted context, we often say that a tuple (even infinite) is contained in $acl_M(A)$ if each element of the tuple is.

As an example, let the structure K be an algebraically closed field of characteristic 0 in the language of rings. Let $A \subseteq K$, and let k be the subfield of K generated by A. Then $b \in \mathrm{acl}_K(A)$ if and only if it is an element of the algebraic closure of k in the algebraic sense. Similarly, $b \in \mathrm{dcl}_K(A)$ if and only if $b \in k$. These both depend on quantifier elimination, and are not completely trivial. (See the author's article Model Theory of Algebraically Closed Fields, in [Be98].)

Lemma 1.2.22. *Assume $M = \overline{M}$, $A \subseteq \overline{M}$ is small, and $b \in M$.*

1. *$b \in \mathrm{acl}_M(A)$ if and only if $\{f(b) : f \in Aut(\overline{M}/A)\}$ is finite.*
2. *$b \in \mathrm{dcl}_M(A)$ if and only if $f(b) = b$ for all $f \in Aut(\overline{M}/A)$.*

Proof.

1. Suppose $b \in \mathrm{acl}_M(A)$ with witness $\exists^{\leqslant k}\varphi(x)$. Then the set defined by $\varphi(x)$ has at most k elements, so is finite. Hence by Lemma 1.2.11, the set $\{m : M \models \varphi(m)\}$ is $\mathrm{Aut}(\overline{M}/A)$-invariant, so $\{f(b) : f \in \mathrm{Aut}(\overline{M}/A)\} \subseteq \{m : M \models \varphi(m)\}$, therefore it is finite as desired.
   Suppose $\{f(b) : f \in \mathrm{Aut}(\overline{M}/A)\}$ is finite. Since the composition of automorphisms is an automorphism, this set is $\mathrm{Aut}(\overline{M}/A)$-invariant. Therefore, by Lemma 1.2.11, it is definable by some formula $\varphi(x)$. Furthermore, $M \models \exists^{\leqslant k}x\varphi(x)$ for some k since the set defined by $\varphi(x)$ is finite. Hence $b \in \mathrm{acl}_M(A)$.
2. Suppose $b \in \mathrm{dcl}_M(A)$ with witness $\exists^{\leqslant 1}\varphi(x)$. Then $\{b\} = \{m \in \overline{M} : \varphi(m)\}$ is an A-definable set, so by Lemma 1.2.11 it is $\mathrm{Aut}(\overline{M}/A)$-invariant. But any singleton which is invariant must be fixed pointwise, so $f(b) = b$ for all $f \in \mathrm{Aut}(\overline{M}/A)$, as desired.
   Suppose $f(b) = b$ for all $f \in \mathrm{Aut}(\overline{M}/A)$. Then $\{b\}$ is $\mathrm{Aut}(\overline{M}/A)$-invariant, so by Lemma 1.2.11 it is definable over A via some formula $\varphi(x)$. Moreover, $M \models \exists^{\leqslant 1}x\varphi(x)$ since $\varphi(x)$ defines a singleton, so $b \in \mathrm{dcl}_{\overline{M}}(A)$.

□

Note the similarity of this lemma to Lemmas 1.2.11 and 1.2.13. In fact, the machinery of T^{eq} could be used to obtain these lemmas as a direct consequence of Lemma 1.2.22.

The first motivation to develop T^{eq} is to deal with quotient objects without leaving the context of first order logic. That is, if E is some definable equivalence relation on some definable set X, we want to view X/E as a definable set.

Let L be a 1-sorted language and let T be a (complete) L-theory. We shall build a many-sorted language L^{eq} and an L^{eq}-theory T^{eq}. We will ensure that in some natural sense, L^{eq} contains L and T^{eq} contains T.

First we define L^{eq}. Consider the set of L-formulas $\varphi(\bar{x}, \bar{y})$, up to equivalence, such that T implies that φ is an equivalence relation (i.e. has reflexivity, transitivity, and symmetry). For each such φ, define s_φ to be a new sort in L^{eq}. Of particular importance is $s_=$, the sort given by the formula "$x = y$." This sort $s_=$ will yield, in each model of T^{eq}, a model of T.

Also define f_φ to be a function symbol with domain sort $s_=^n$ (where φ has n free variables) and codomain sort s_φ.

For each m-place relation symbol $R \in L$, make R^{eq} an m-place relation symbol in L^{eq} on $s_=$. Likewise for all constant and function symbols in L. Finally, we put an equality symbol $=_\varphi$ on each sort.

Remark 1.2.23. Let N be an L^{eq} structure. Then N has interpretations $s_\varphi(N)$ of each sort s_φ and $f_\varphi(N) : s_=(N)^{n_{f_\varphi}} \to s_\varphi(N)$ of each function symbol f_φ. Additionally, N will contain an L-structure consisting of $s_=$ and interpretations of the symbols of L inside $s_=$ as above (e.g. R^{eq} for R).

Definition 1.2.24. T^{eq} is the L^{eq} theory which is axiomatized by the following:

1. T, where the quantifiers in the formulas of T now range over the sort $s_=$.
2. For each suitable L-formula $\varphi(\bar{x}, \bar{y})$, the axiom $(\forall_{s_=} \bar{x})\ (\forall_{s_=} \bar{y})\ (\varphi(\bar{x}, \bar{y}) \leftrightarrow f_\varphi(\bar{x}) = f_\varphi(\bar{y}))$.
3. For each L-formula φ, the axiom $(\forall_{s_\varphi} y)\ (\exists_{s_=} \bar{x})\ f_\varphi(\bar{x}) = y$.

Note that axioms 2 and 3 simply state that f_φ is the quotient function for the equivalence relation given by φ. Also as the notation suggests, in many-sorted logic any quantifier ranges over a sort.

Definition 1.2.25. Let $M \models T$. Then M^{eq} is the L^{eq} structure such that $s_=(M^{eq}) = M$, and for each suitable L-formula $\varphi(\bar{x}, \bar{y})$ in $2n$ variables, the sort $s_\varphi(M^{eq})$ is equal to M^n/E, where E is the equivalence relation defined by $\varphi(\bar{x}, \bar{y})$, and $f_\varphi(M^{eq})(\bar{b}) = \bar{b}/E$.

One can now easily verify that $M^{eq} \models T^{eq}$. Moreover, passing from T to T^{eq} is a canonical operation, in the following sense:

Lemma 1.2.26.

1. *For any $N \models T^{eq}$, there is an $M \models T$ such that $N \cong M^{eq}$.*
2. *Suppose $M, N \models T$ are isomorphic, and let $h : M \cong N$. Then h extends uniquely to $h^{eq} : M^{eq} \cong N^{eq}$.*
3. *T^{eq} is a complete L^{eq}-theory.*
4. *Suppose $M, N \models T$ and let $\bar{a} \in M, \bar{b} \in N$ with $\text{tp}_M(\bar{a}) = \text{tp}_N(\bar{b})$. Then $\text{tp}_{M^{eq}}(\bar{a}) = \text{tp}_{N^{eq}}(\bar{b})$.*

Proof.

1. Let $N \models T^{eq}$, we can take $M = s_=(N)$ with its canonical L-structure.
2. Suppose $M, N \models T$ are isomorphic, and let $h : M \cong N$. Let $h^{eq} : M^{eq} \to N^{eq}$ be defined as $h^{eq}(f_\varphi(M^{eq})(\bar{b})) = f_\varphi(N^{eq})(h(\bar{b}))$ for each L-formula φ. This defines a function on M^{eq}, because $f_\varphi(M^{eq})$ is surjective by the T^{eq} axioms. Moreover h^{eq} is well-defined, bijective by the construction of M^{eq} and N^{eq}, and an isomorphism by definition.
3. Let $M, N \models T^{eq}$, we want to show that they are elementarily equivalent. Assuming the generalized continuum hypothesis (GCH), there are $M', N' \models T^{eq}$ which are λ saturated of size λ, for some large λ, with $M \preceq M'$ and $N \preceq N'$. Since we want to show elementary equivalence, we can replace M, N with M' and N'. By 1, we have $M = M_0^{eq}, N = N_0^{eq}$ for some $M_0, N_0 \models T$. Furthermore, M_0, N_0 are λ-saturated of size λ. By assumption, T is complete, so $M_0 \equiv N_0$, and therefore $M_0 \cong N_0$. By 2, $M \cong N$, and therefore $M \equiv N$.
4. Similar proof to 3. We may assume $M = N$ is saturated. There is an automorphism h taking $\bar{a}$ to $\bar{b}$. But h extends to an automorphism of M^{eq} by 1, so $\bar{a}$ and $\bar{b}$ have the same type in M^{eq}.

□

Corollary 1.2.27. *Consider the Stone space $S_{(s_=)^n}(T^{eq})$. The forgetful map $\pi : S_{(s_=)^n}(T^{eq}) \to S_n(T)$ is a homeomorphism.*

Proof. Observe that it is continuous and surjective. By part 4 of the previous lemma, it is injective. Any continuous bijection from a compact space to a Hausdorff space is a homeomorphism. □

Proposition 1.2.28. *Let $\varphi(x_1, \ldots, x_k)$ be an L^{eq} formula, where x_i is of sort S_{E_i}. There is an L-formula $\psi(\bar{y_1}, \ldots, \bar{y_k})$ such that:*

$$T^{eq} \models (\forall \bar{y_1}) \ldots (\forall \bar{y_k}) (\psi(\bar{y_1}, \ldots, \bar{y_k}) \leftrightarrow \varphi(f_{E_1}(\bar{y_1}), \ldots, f_{E_k}(\bar{y_k}))).$$

Proof. Let n be the length of $\bar{y_1} \ldots \bar{y_k}$. Consider the set $\pi(\varphi(f_{E_1}(\bar{y_1}), \ldots, f_{E_k}(\bar{y_k})))$, it is a clopen subset of $S_n(T)$ by the previous corollary, hence equal to $\psi(\bar{y_1}, \ldots, \bar{y_k})$ for some formula ψ. One easily checks this formula is the one we are looking for. □

Corollary 1.2.29.

1. *Let $M, N \models T$, and let $h : M \to N$ be an elementary embedding. Then $h^{eq} : M^{eq} \to N^{eq}$ (defined as was done earlier) is also an elementary embedding.*
2. *$\bar{M}^{eq}$ is also κ-saturated.*

Remark 1.2.30. For $M \models T$, a definable set $X \subseteq M^n$ can be viewed as an element of M^{eq} as follows. Suppose X is defined in M by $\varphi(\bar{x}, \bar{a})$ where $\bar{a} \in M$. Consider the equivalence relation E_ψ defined by $\psi(\bar{y_1}, \bar{y_2}) = (\forall \bar{x}) (\varphi(\bar{x}, \bar{y_1}) \leftrightarrow \varphi(\bar{x}, \bar{y_2}))$, and consider $c = \bar{a}/E_\psi = f_\psi(\bar{a}) \in M^{eq}$. Then X is defined in M^{eq} by $\chi(\bar{x}, c) = (\exists \bar{y}) (\varphi(\bar{x}, \bar{y}) \wedge f_\psi(\bar{y}) = c)$. Moreover, if $c' \in S_\psi(M^{eq})$, and $M^{eq} \models (\forall \bar{x}) (\chi(\bar{x}, c) \leftrightarrow \chi(\bar{x}, c'))$, then $c = c'$. To see this, let $c' = f_\psi(\bar{a}')$, and let X' be defined in M by $\varphi(\bar{x}, \bar{a}')$. Then X' is defined in M^{eq} by $\chi(\bar{x}, c')$, so we have that $X = X'$ (in M^{eq}). And then $X = X'$ (in M), so $c = f_\psi(\bar{a}) = f_\psi(\bar{a}') = c'$.

Definition 1.2.31. With the above considerations in mind, given $M \models T$ and a definable set $X \subseteq M^n$, we call such a $c \in M^{eq}$ a code for X.

Remark 1.2.32. An automorphism of $\bar{M}^{eq}$ fixes a definable set X setwise if and only if it fixes a code for X. Note that the choice of a code for X will depend on the L-formula $\varphi(\bar{x}, \bar{y})$ in Remark 1.2.30, and so codes are not necessarily unique, although interdefinable.

Definition 1.2.33. Let $A \subseteq M \models T$. Then $\mathrm{acl}^{eq}(A) = \{c \in M^{eq} : c \in \mathrm{acl}_{M^{eq}}(A)\}$, and $\mathrm{dcl}^{eq}(A)$ is defined similarly.

Remark 1.2.34. Suppose $A \subseteq M \prec N$. Then $\mathrm{acl}_{N^{eq}}(A), \mathrm{dcl}_{N^{eq}}(A) \subseteq M^{eq}$, so this notation is unambiguous.

Lemma 1.2.35. *Let $M \models T$, a definable subset X of M^n, and $A \subseteq M$. Then X is almost over A if and only if X is definable in M^{eq} by a formula with parameters in $\mathrm{acl}^{eq}(A)$.*

Proof. We can work in $\bar{M}$. Let c be a code for X. From Lemma 1.2.13, X is almost over A if and only if $|\{\sigma(X) : \sigma \in \mathrm{Aut}(\bar{M}/A)\}| < \omega$, if and only if $|\{\sigma(c) : \sigma \in \mathrm{Aut}(\bar{M}/A)\}| < \omega$, that is $c \in \mathrm{acl}^{eq}(A)$. □

Definition 1.2.36. Let $\bar{a}, \bar{b} \in \bar{M}$ have length n. Let $A \subseteq \bar{M}$. Then $\bar{a}, \bar{b}$ have the same strong type over A (written as $\mathrm{stp}_{\bar{M}}(\bar{a}/A) = \mathrm{stp}_{\bar{M}}(\bar{b}/A)$) if $E(\bar{a}, \bar{b})$ for any finite equivalence relation defined over A.

Remark 1.2.37. If $\varphi(\bar{x})$ is a formula over A, then it defines an equivalence relation with two classes by the formula $E(\bar{x}_1, \bar{x}_2)$: $(\varphi(\bar{x}_1) \wedge \varphi(\bar{x}_2)) \vee (\neg\varphi(\bar{x}_1) \wedge \neg\varphi(\bar{x}_2))$. Hence, strong types refine types.

In the context of models of the complete theory T, when we talk about type of a tuple over a set, it is in the sense of the ambient saturated model $\bar{M}$.

Lemma 1.2.38. *If $\bar{a}$ is a tuple (from $\bar{M}$) and M is a model (elementary substructure of $\bar{M}$), then* $\mathrm{tp}(\bar{a}/M)$ *implies* $\mathrm{stp}(\bar{a}/M)$.

Proof. Let E be an equivalence relation with finitely many classes, defined over M, and $\bar{b}$ another realization of $\mathrm{tp}_{\bar{M}}(\bar{a}/M)$, we want to show $E(\bar{a}, \bar{b})$. Since E has only finitely many classes, and M is a model, there are representatives in M, say $\bar{e}_1, \cdots, \bar{e}_n$ of each E-class in $\bar{M}$. Hence we must have $E(\bar{a}, \bar{e}_i)$ for some i, and therefore also $E(\bar{b}, \bar{e}_i)$, which yields $E(\bar{a}, \bar{b})$, which we wanted. □

Lemma 1.2.39. *Let $A \subseteq M \models T$, and let $\bar{a}, \bar{b} \in M$. Then the following are equivalent:*

1. $\mathrm{stp}(\bar{a}/A) = \mathrm{stp}(\bar{b}/A)$
2. *$\bar{a}, \bar{b}$ satisfy the same formulas almost over A*
3. $\mathrm{tp}(\bar{a}/\mathrm{acl}^{eq}(A)) = \mathrm{tp}(\bar{b}/\mathrm{acl}^{eq}(A))$.

Proof. 2 ⇔ 3 is a direct consequence of Lemma 1.2.35, so we just need to prove 1 ⇔ 2.

First, let's do the left to right implication. Assume $\bar{a}, \bar{b} \in M$ and $\mathrm{stp}(\bar{a}/A) = \mathrm{stp}(\bar{b}/A)$. Let X be definable almost over A, we want to show that $\bar{a} \in X$ if and only if $\bar{b} \in X$. By symmetry, it is enough to do one direction. So assume $\bar{a} \in X$.

Since X is almost over A, there is an A-definable equivalence relation E with finitely many classes, and $\bar{c}_1, \cdots, \bar{c}_n$ such that for all $\bar{x} \in M$, we have

$\bar{x} \in X$ if and only if $M \models E(\bar{x}, \bar{c}_1) \vee \cdots \vee E(\bar{x}, \bar{c}_n)$. Hence $E(\bar{a}, \bar{c}_i)$ for some i, so by assumption $E(\bar{b}, \bar{c}_i)$. Therefore $\bar{b} \in X$.

Now for the right to left direction, suppose $\bar{a}, \bar{b}$ satisfy the same almost over A formulas. Let E be an A-definable equivalence relation with finitely many classes, we want to show $E(\bar{a}, \bar{b})$. The set $X = \{\bar{x} \in M, E(\bar{x}, \bar{a})\}$ is definable almost over A. But $\bar{a} \in X$, so $\bar{b} \in X$, hence $E(\bar{a}, \bar{b})$. □

Definition 1.2.40.

1. T has elimination of imaginaries (EI) if, for any model $M \models T$ and $e \in M^{eq}$, there is a $\bar{c} \in M$ such that $e \in \mathrm{dcl}^{eq}(\bar{c})$ and $\bar{c} \in \mathrm{dcl}^{eq}(e)$.
2. T has weak elimination of imaginaries if, same as above, except $\bar{c} \in \mathrm{acl}^{eq}(e)$.
3. T has geometric elimination of imaginaries if, same as above, except $e \in \mathrm{acl}^{eq}(\bar{c})$ and $\bar{c} \in \mathrm{acl}^{eq}(e)$.

Note that in particular, elimination of imaginaries imply the existence of *real* codes for definable sets. Moreover, we have the following characterization:

Proposition 1.2.41. *The following are equivalent:*

1. *T has EI.*
2. *For some model $M \models T$, we have that for any $\emptyset$-definable equivalence relation E, there is a partition of M^n into $\emptyset$-definable sets $Y_1, \ldots, Y_r$, and for each $i = 1, \ldots, r$ a $\emptyset$-definable $f_i : Y_i \to M^{k_i}$ where $k_i \geqslant 1$ such that for each $i = 1, \ldots, r$, for all $\bar{b_1}, \bar{b_2} \in Y_i$, we have $E(\bar{b_1}, \bar{b_2})$ iff $f_i(\bar{b_1}) = f_i(\bar{b_2})$.*
3. *For any model $M \models T$, we have that for any $\emptyset$-definable equivalence relation E, there is a partition of M^n into $\emptyset$-definable sets $Y_1, \ldots, Y_r$, and for each $i = 1, \ldots, r$ a $\emptyset$-definable $f_i : Y_i \to M^{k_i}$ where $k_i \geqslant 1$ such that for each $i = 1, \ldots, r$, for all $\bar{b_1}, \bar{b_2} \in Y_i$, we have $E(\bar{b_1}, \bar{b_2})$ iff $f_i(\bar{b_1}) = f_i(\bar{b_2})$.*
4. *For any model $M \models T$, and any definable $X \subseteq M^n$ there is an L-formula $\varphi(\bar{x}, \bar{y})$ and $\bar{b} \in M$ such that X is defined by $\varphi(x, \bar{b})$ and for all $\bar{b}' \in M$ if X is defined by $\varphi(\bar{x}, \bar{b}')$ then $\bar{b} = \bar{b}'$. (We call such a $\bar{b}$ a* real *code for X.)*

Proof. Note that properties in 2 and 3 concern only $\emptyset$-definable relations and functions. Hence, if it is true in some model, it is true in any model, so 2 and 3 are equivalent. As a consequence, we can and will work in our saturated model $\bar{M}$ (or $\bar{M}^{eq}$). When we talk about *real* tuples we mean finite tuples from the "home sort" $\bar{M}$.

We start with 1 $\Rightarrow$ 2, and therefore consider some $\varnothing$-definable equivalence relation E. Let $\pi_E : S^n_= \to S_E$ be the canonical definable quotient map. Let $e \in S_E$. By assumption, there is $\bar{c} \in \overline{M}^k$ (some k) such that e and $\bar{c}$ are interdefinable. In other words, there is a L^{eq}-formula $\varphi_e(w, \bar{y})$ over $\varnothing$ such that $\varphi_e(e, \bar{c})$. Moreover, $\bar{c}$ is the unique tuple $\bar{y}$ such that $\models \varphi_e(e, \bar{y})$, and e is the unique w such that $\models \varphi_e(x, \bar{c})$.

Let $X_e = \{\bar{x} \in \overline{M}^k :\models (\exists! \bar{y} \varphi_e(\pi_E(x), \bar{y})) \wedge ((\forall \bar{z}(E(\bar{x}, \bar{z})) \leftrightarrow (\forall \bar{y}(\varphi_e (\pi_E(\bar{x}), \bar{y})) \leftrightarrow (\varphi_e(\pi_E(\bar{z}), \bar{y})))))$. This means that φ_e defines a function on X_e, and that this function separates E-classes.

We claim that $\pi_E^{-1}(\{e\}) \subseteq X_e$. Let $\bar{a} \in \pi^{-1}(\{e\})$, so $\bar{c}$ is the unique realization of $\varphi_e(\pi_E(\bar{a}), \bar{y})$. Suppose $E(\bar{a}, \bar{b})$, then $\pi_E(\bar{a}) = e = \pi_E(\bar{b})$, so obviously $\varphi_e(\pi_E(\bar{a}), \bar{y})$ and $\varphi_e(\pi_E(\bar{b}), \bar{y})$ define the same set. On the other hand suppose that $\models \forall \bar{y}(\varphi_e(\pi_E(\bar{a}), \bar{y}) \leftrightarrow \varphi_e(\pi_E(\bar{b}), \bar{y}))$, for some $\bar{b}$. By assumption, we have $\varphi_e(\pi_E(\bar{a}), \bar{c})$, hence $\varphi_e(\pi_e(\bar{b}), \bar{c})$. But by definition of φ_e, this implies that $e = \pi_E(\bar{b})$, and by definition of π_E, this yields $E(\bar{a}, \bar{b})$.

Since each X_e contains $\pi^{-1}(\{e\})$, we get $\overline{M}^n = \bigcup_{e \in \pi_E(\overline{M}^n)} X_e$, and by compactness, there are $e_1, \ldots, e_l$ such that $\overline{M}^n = \bigcup_{i=1}^{l} X_{e_i}$. We almost have what we wanted, by considering $f_i = \varphi_{e_i} \circ \pi_E$ on each X_{e_i}. However, the X_{e_i} are not disjoint.

But we can consider $Y_1, \cdots, Y_r$ to be the atoms of the boolean algebra generated by the X_i. These are disjoint, and we can pick, for each Y_j, an appropriate f_i, to get the result.

We now prove 3 $\Rightarrow$ 4. Let X be defined by $\varphi(x, \bar{a})$. Consider the $\varnothing$-definable equivalence relation $E(\bar{y}, \bar{z}) : \forall \bar{x}(\varphi(\bar{x}, \bar{y}) \leftrightarrow \varphi(\bar{x}, \bar{z}))$. Let Y_i and f_i be as in 2, and say $\bar{a} \in Y_1$, and let $\bar{b} = f_1(\bar{a})$. Then $\exists \bar{y}(f_1(\bar{y}) = \bar{b} \wedge \varphi(x, \bar{y}))$ defines X, call this formula $\psi(x, \bar{b})$.

We have to show that $\bar{b}$ is unique such. Let $\bar{b}'$ be such that $\exists \bar{y}(f_1(\bar{y}) = \bar{b}' \wedge \varphi(x, \bar{y}))$ also defines X, and let $\bar{a}_0$ be such that $f_1(\bar{a}_0) = \bar{b}'$ and $\varphi(\bar{x}, \bar{a}_0)$ defines X. Then $\bar{a}_0 E \bar{a}$, which implies $\bar{b}' = f_1(\bar{a}_0) = f_1(\bar{a}) = \bar{b}$.

Finally, we need to prove 4 $\Rightarrow$ 1. Let $e \in \mathcal{M}^{eq}$, then $e = \pi_E(\bar{a})$, for some $\bar{a} \in \overline{M}^n$ and some $\varnothing$-definable equivalence relation E.

The set $X = \{\bar{x} \in \overline{M}^n :\models E(\bar{x}, \bar{a})\}$ has a code $\bar{b} \in \overline{M}^k$, so that $X = \psi(\overline{M}^n, \bar{b})$. Then an automorphism of $\bar{M}^{eq}$ fixes e iff it fixes the set defined by $E(\bar{x}, \bar{a})$ iff it fixes the tuple $\bar{b}$. So e and $\bar{b}$ are interdefinable. $\square$

It would be nice to have $r = 1$ in the above proposition. Here is a condition for this to be true:

Proposition 1.2.42. *Suppose T eliminates imaginaries and* $\mathrm{dcl}(\emptyset)$ *has at least two elements. Then for any model M of T and $\emptyset$-definable equivalence relation E on M^n there is a $\emptyset$-definable function f from M^n to M^k for some k such that $E(\bar{x}, \bar{y})$ is equivalent to $f(\bar{x}) = f(\bar{y})$ in M (or in T).*

Proof. Suppose that $\mathrm{dcl}(\emptyset)$ contains two elements a and b. Let Y_i, f_i be as in condition 2 of Proposition 1.2.41. By the proof there we may assume that each Y_i is a union of E-classes. Using a and b, we can find some number k and functions $g_i : \overline{M}^{k_i} \to \overline{M}^k$ such that the $g_i(\overline{M}^{k_i})$ are pairwise disjoint. We can check that the $\emptyset$-definable function $f : \overline{M}^n \to \overline{M}^k$ sending $y \in Y_i$ to $g_i(f_i(y))$ has all the required properties. □

Remark 1.2.43. Elimination of imaginaries also makes sense for many sorted theories. Bearing this in mind, we will now give a lemma.

Lemma 1.2.44 (Assume T 1-sorted). *T^{eq} has elimination of imaginaries.*

Proof. We prove a strong version of (2) in Proposition 1.2.41. Let E' be a $\emptyset$-definable equivalence relation on a sort s_E in some model M^{eq} of T^{eq}. By Proposition 1.2.28, there is an L-formula $\psi(\overline{y}_1, \overline{y}_2)$ ($\overline{y}_i$ the appropriate length) such that for all $\overline{a}_1, \overline{a}_2 \in M$, $M \models \psi(\overline{a}_1, \overline{a}_2)$ if and only if $M^{eq} \models E'(f_E(\overline{a}_1), f_E(\overline{a}_2))$. So $\psi(\overline{y}_1, \overline{y}_2)$ is an L-formula defining an equivalence relation on M^k for the suitable length k. Consider the map h, taking $e \in S_E$ to $f_\psi(\overline{a}) \in S_\psi$ for any $\overline{a} \in M^k$ such that $f_E(\overline{a}) = e$. Suppose $f_E(\overline{a}) = e = f_E(\overline{a'})$, we easily see that $f_\psi(\overline{a}) = f_\psi(\overline{a'})$, hence the map h is well defined, and satisfies (2) of 1.2.41. □

In so-called applied model theory, a substantial amount of work has been done on proving some theorems about relative elimination of imaginaries. That is, given a theory T, are there some natural imaginary sorts one can add to T so that the resulting theory has elimination of imaginaries? The case of valued fields is interesting, with applications.

1.2.4 *Examples and counterexamples*

Example 1.2.45. The theory of an infinite set has weak elimination of imaginaries but not full elimination of imaginaries.

Proof. First, we show that T has weak elimination of imaginaries. Let M be an infinite set and let $e \in M^{eq}$ be an imaginary element. Suppose that $e = \lceil X \rceil$ for some definable set X. Let $A \subset M$ be a finite set over which X is definable. Consider the set

$$\hat{A} := \bigcap_{\substack{\sigma \in \mathrm{Aut}(M) \\ \sigma(X)=X}} \sigma(A).$$

Since A is finite, there are $\sigma_1, \ldots, \sigma_n$ such that $\hat{A} = \cap_i \sigma_i(A)$. Observe that X is definable over $\hat{A}$; since automorphisms of M are just permutations, for any finite sets B_1 and B_2, $\mathrm{Aut}(M/B_1 \cap B_2)$ is generated by $\mathrm{Aut}(M/B_1)$ and $\mathrm{Aut}(M/B_2)$. Therefore if X is definable over B_1 and B_2, then X is definable over $B_1 \cap B_2$. So $e \in dcl(\hat{A})$. It remains to show that $\hat{A} \subseteq \mathrm{acl}^{eq}(e)$. Let $a \in \hat{A}$. For any $\sigma \in \mathrm{Aut}(M^{eq})$ fixing e, we have that $\sigma(X) = X$ and so $\sigma(\hat{A}) = \hat{A}$, i.e. $\sigma(a) \in \hat{A}$. But $\hat{A}$ is finite, so a has only finitely many images under automorphisms fixing e so $a \in acl(e)$.

To see that T does not have full elimination of imaginaries, just observe that if $a \neq b \in M \models T$, then there is no code for the definable set $\{a, b\}$: for any finite tuple $\bar{c}$ there is an automorphism of M which fixes $\{a, b\}$ as a set but does not fix the tuple $\bar{c}$. □

Example 1.2.46. Let $T = Th(M, <, ...)$ where $<$ is a total well-ordering (such as True Arithmetic). Then T has elimination of imaginaries.

Proof. Every definable set has a first element since $<$ is a total well-ordering. We verify (2) in 1.2.41. Let E be a $\emptyset$-definable equivalence relation on M^n. Let $f : M^n \to M^n$ such that for any $\overline{a}$, $f(\overline{a})$ is the least element of the E-class of $\overline{a}$, with respect to the lexicographic ordering. Notice that f is $\emptyset$-definable, and for all $\bar{a}, \bar{b}$, we have $f(\bar{a}) = f(\bar{b})$ if and only if $E(\bar{a}, \bar{b})$. □

Therefore, one should not think of elimination of imaginaries as measuring the complexity of a theory.

Definition 1.2.47. A (one-sorted) theory T is strongly minimal if for any model M of T and any definable set $X \subset M$, either X is finite or $M \backslash X$ is finite.

Example 1.2.48. The following theories are strongly minimal:

- the theory of equality
- the theory of an algebraically closed field
- $\mathrm{Th}(\mathbb{Q}, +)$.

From the point of view of model theory, strongly minimal theories are very well understood, and in particular, we have:

Lemma 1.2.49. *Let T be strongly minimal and* $\mathrm{acl}(\emptyset)$ *be infinite (in some, any model). Then T has weak elimination of imaginaries.*

Proof. Fix a model M. Let $e \in M^{eq}$, so $e = \overline{a}/E$ for some $\overline{a} = (a_1, ..., a_n)$ and E some $\emptyset$-definable equivalence relation on n-tuples. Let $A = \mathrm{acl}(e) \cap M$ (i.e. those elements of the "home sort" M which are in the algebraic closure of e in the structure M^{eq}). A is infinite as it contains $\mathrm{acl}(\emptyset)$.

We first prove that there exists some tuple $\overline{b}$ of elements of A such that $E(\overline{a}, \overline{b})$. Let $X_1 = \{y_1 \in M : M \models \exists y_2, ..., y_n(\overline{y}E\overline{a})\}$, which is definable over e. If X_1 is finite then it is a subset of A. Otherwise, X_1 is cofinite, hence meets the infinite set A. Either way, $X_1 \cap A \neq \emptyset$ and choose $b_1 \in X_1 \cap A$.

Now let $X_2 = \{y_2 \in M : M \models \exists y_3, ..., y_n(b_1 y_2 ... y_n E\overline{a})\}$. We remark that $X_2 \neq \emptyset$ since $b_1 \in X_1$. Now, X_2 is either finite or cofinite since T is strongly minimal. By the same argument above, we may find $b_2 \in X_2 \cap A$. Then, repeating this process, we may find the tuple $\overline{b}$ from A. Therefore, $\overline{b} \in \mathrm{acl}(e)$. And of course $e \in dcl(\overline{b})$. □

Example 1.2.50. The theory ACF_p algebraically closed fields of characteristic p, has elimination of imaginaries, for any p (zero or a prime).

Proof. By Lemma 1.2.49, ACF_p has weak elimination of imaginaries. Therefore, it suffices to show that every finite set of n-tuples can be coded. Let $\bar{c}_1, \ldots, \bar{c}_m$ be n-tuples from the algebraically closed field K. Let $P(Z, X_1, \ldots, X_n)$ be the polynomial $(Z + c_{11}X_1 + \cdots + c_{1n}X_n) \ldots (Z + c_{m1}X_1 + \cdots + c_{mn}X_n)$.

Let $\bar{d}$ be the tuple of coefficients of P. Let e be the imaginary $\{\bar{c}_1, \ldots, \bar{c}_m\}$, then e is interdefinable with the tuple $\bar{d}$. □

Example 1.2.51. The theory of real closed fields has elimination of imaginaries. To see this, we prove that $T = RCF$ has definable choice.

Definition 1.2.52. A theory T has definable choice if for any L-formula $\varphi(\bar{x}, \bar{y})$ and any model M of T, there is a $\emptyset$-definable partial function f_φ from $\bar{y}$ tuples in M to $\bar{x}$-tuples in M such that:

if $M \models \exists \bar{x}\varphi(\bar{x}, \bar{a})$ then $M \models \varphi(f_\varphi(\bar{a}), \bar{a}))$, and if $\varphi(\bar{x}, \bar{b})$ and $\varphi(\bar{x}, \bar{c})$ are equivalent in M then $f_\varphi(\bar{b}) = f_\varphi(\bar{c})$.

We can then use the general property:

Proposition 1.2.53. *If T has definable choice, then it eliminates imaginaries.*

Proof. Work in any model of T.

Let E be an equivalence relation, definable over the empty set. Consider the $\emptyset$-definable function f_E.

Then for any $\bar{a}, \bar{b}$ we see from the definitions that $E(\bar{a}, \bar{b})$ if and only if $f_E(\bar{a}) = f_E(\bar{b})$, which yields elimination of imaginaries by Lemma 1.2.41. □

Proposition 1.2.54. *The theory of real closed fields has definable choice.*

Proof. Fix a real-closed field K and L-formula $\varphi(\bar{x}, \bar{y})$. We proceed by induction on $|\bar{x}|$.

$|\bar{x}| = 1$: By quantifier elimination (or o-minimality), for any $\bar{b}$, the set X defined by $\varphi(x, \bar{b})$ is a finite union of intervals (we will assume points are intervals as well). Let I be the left-most interval of X. We define $f(\bar{b})$ in cases depending on the shape of I:

- if $I = \{a\}$ is a singleton, set $f(\bar{b}) = a$,
- if $I = \mathcal{R}$, set $f(\bar{b}) = 0$,
- if the interior of I is of the form $(c, +\infty)$, set $f(\bar{b}) = c + 1$,
- if the interior of I is of the form $(-\infty, c)$, set $f(\bar{b}) = c - 1$,
- if the interior of I is of the form (c, d), then set $f(\bar{b}) = \frac{c+d}{2}$.

Note that f is $\emptyset$-definable and its value at $\bar{b}$ depends only the set X defined by $\varphi(x, \bar{b})$.

Now consider the case where $|\bar{x}| = n + 1$: Write $\bar{x} = (x_0, \bar{x}_1)$, where $|\bar{x}_1| = n$. Let $\psi(\bar{x}_1, \bar{y})$ be $\exists x_0(\varphi(x_0, \bar{x}_1, \bar{y}))$.

By the induction hypothesis, there is a $\emptyset$-definable partial function $f(\bar{y})$ corresponding to the formula $\psi(\bar{x}_1, \bar{y})$.

But also by only the case of $|\bar{x}| = 1$, there is a $\emptyset$-definable partial function $g(\bar{x}_1, \bar{y})$ corresponding to the formula $\varphi(x_0, \bar{x}_1, \bar{y})$.

Then the $\emptyset$-definable partial function $h(\bar{y}) = g(f(\bar{y}), \bar{y})$ works for the formula $\varphi(\bar{x}, \bar{y})$. □

Finally we relate definable choice to Skolem functions.

Definition 1.2.55. T has definable Skolem functions if for each formula $\varphi(\bar{x}, \bar{y})$ there is some definable (over the empty set) function $f_\varphi(\bar{y})$ such that for any model M of T, we have $M \models \forall y((\exists \bar{x}\varphi(\bar{x}, \bar{y})) \to (\varphi(f_\varphi(\bar{y}), \bar{y})))$.

Note that definable choice implies definable Skolem functions, but the converse is false. Indeed, the theory $\mathrm{Th}(\mathbb{Q}_p, +, \times)$ has Skolem functions, but does not have elimination of imaginaries, and hence does not have definable choice.

However we have the rather easy observation which is left to the reader.

Fact 1.2.56. *A theory T has definable choice if and only if T^{eq} has definable Skolem functions.*

The following question was raised in the lectures but we did not explore it and are not sure of the interest.

Question 1. Consider the theories ZF and ZFC. Which completions of these theories have Skolem functions, elimination of imaginaries, or definable choice?

1.3 Stability

Throughout this chapter we will fix a complete theory T in some language L. We will have no problem in working in T^{eq} (that is to say, to assume $T = T^{\mathrm{eq}}$), at least for the general theory. Of course some definitions, such as strong minimality depend on a specific "home sort".

We start with a little bit of history. The origin of stability theory can be traced back to Morley's work in the sixties regarding uncountable categoricity. He proved the following theorem:

Theorem. *Suppose T is a countable theory. Then T is κ-categorical for some $\kappa > \aleph_0$ if and only if T is κ-categorical for all $\kappa > \aleph_0$.*

A key step in the proof of this theorem is to show that, if T is κ-categorical for some $\kappa > \aleph_0$, then T is ω-stable (a property called "totally transcendental" by Morley). This property, ω-stability, is a strong form of stability, defined as follows: for all $n < \omega$ and all countable models $M \models T$, the cardinality of $S_n(M)$ is at most countable (which in turn implies $|S_n(M)| \leqslant |M|$ for *all* $M \models T$).

Assuming ω-stability, additional machinery was developed (such as Morley rank) to deduce the theorem. Subsequently, it was seen that, in fact, a theory T is κ-categorical for some/any $\kappa > \aleph_0$ if and only if T is ω-stable and unidimensional (i.e. any two types are nonorthogonal), a characterisation of $\aleph_1$-categorical theories without any mention of an uncountable cardinals.

Later on, Shelah formulated "test questions" that could help classify first order theories, such as what are the possible spectrum functions of a first order complete (countable) theory T. For a given theory T, the spectrum function is given as

$$I(T,-) : \text{cardinals} \longrightarrow \text{cardinals}$$
$$I(T,\lambda) = \# \text{ of models of } T \text{ of cardinality } \lambda \text{ (up to isomorphism).}$$

In solving this question, he invented stability theory. For instance, if T is unstable, then $I(T,\lambda) = 2^\lambda$ for all $\lambda > \aleph_0$, i.e. T has the greatest possible number of models in each uncountable cardinality. Hence in order to solve the spectrum problem (restricted to uncountable cardinals) one can restrict attention to stable theories. See [She90] for the original work on stability theory, as well as to the solution of the spectrum problem. There will be more discussion in Section 1.3.3.

As we have recently discovered, the notion of stability in the sense of a stable (real-valued) relation $R(x,y)$ appears in Grothendieck's thesis from the early 1950's in functional analysis. See the paper [Gro52]. Our exposition of local stability will be influenced by his results and even proofs.

1.3.1 *Local stability*

We begin by recalling that, for an L-formula φ, to write φ as $\varphi(x_1,\dots,x_n)$ means that the free variables of φ are among $x_1,\dots,x_n$ (which we assume are distinct). Similarly, writing φ as $\varphi(\overline{x},\overline{y})$ means $\overline{x},\overline{y}$ form a partition of the free variables of φ.

Definition 1.3.1.

(i) Let $M \models T$. We say $\varphi(\overline{x},\overline{y})$ is *stable in* M if it is *not* the case that there are $\overline{a}_i, \overline{b}_i$ in M, for $i < \omega$, such that *either*, for all $i,j < \omega$, $M \models \varphi(\overline{a}_i,\overline{b}_j)$ if and only if $i \leqslant j$, *or*, for all $i,j<\omega$, $M \models \neg\varphi(\overline{a}_i,\overline{b}_j)$ if and only if $i \leqslant j$.

(ii) We say $\varphi(\overline{x},\overline{y})$ is *stable* (for T) if it is stable in M for all $M \models T$.

(iii) Finally, we say T is *stable* if every L-formula $\varphi(\overline{x},\overline{y})$ is stable (for T).

Remark 1.3.2. A formula $\varphi(\overline{x},\overline{y})$ is stable for T if and only if it is *not* the case that there are $\overline{a}_i, \overline{b}_i$ in the monster model $\overline{M}$, $i<\omega$, such that $\overline{M} \models \varphi(\overline{a}_i,\overline{b}_i)$ if and only if $i \leqslant j$ for all $i,j<\omega$.

Before we go on, a remark on notation. For simplicity, from now onwards we will write tuples simply as x and a, instead of the more cumbersome $\overline{x}$ and $\overline{a}$. Thus, in general, when say we write $\varphi(x, y)$, we understand x and y not as single variables but as tuples (possibly of length 1) of variables. On the other hand in $\overline{M}^{eq}$ a finite tuple of elements (even from different sorts) can be identified with an element of another sort, so this notation is not too problematic.

We start by observing a few elementary facts about stable formulas:

Lemma 1.3.3.

(i) Suppose $\varphi(x, y)$ and $\psi(x, z)$ are stable for T. Then $\neg\varphi(x, y)$, $(\varphi \vee \psi)(x, yz)$ and $(\varphi \wedge \psi)(x, yz)$ are also stable.

(ii) Given $\varphi(x, y)$, let $\varphi^*(y, x)$ be $\varphi(x, y)$. Then, $\varphi(x, y)$ stable for T implies $\varphi^*(y, x)$ is stable for T as well.

(iii) The formula $\varphi(x, y)$ is stable for T if and only if there is $n < \omega$ such that $\varphi(x, y)$ is n-stable: it is not *the case that there are a_i, b_i (in $\overline{M}$ or in some/any $M \models T$), for $i \leqslant n$, such that $\models \varphi(a_i, b_i)$ if and only if $i \leqslant j$ for all $i, j \leqslant n$.*

(iv) There are T, $M \models T$ and $\varphi(x, y)$ such that $\varphi(x, y)$ is stable in M but it is not stable for T.

In particular, as a consequence of the lemma, notice that (iii) implies that stability is expressed by a sentence of the theory.

Proof.

(i) The case of negation is immediate from Definition 1.3.1(i). The rest is left to the reader.

(ii) Suppose $\varphi^*(y, x)$ is not stable, so by (i) $\neg\varphi^*(y, x)$ is also unstable. Let a_i, b_i be witnesses in $\overline{M}$ of the latter. Then, $a'_i = b_i$ and $b'_i = a_{i+1}$, $i < \omega$, witness the instability of $\varphi(x, y)$, as $j + 1 > i$ holds if and only if $i \leqslant j$. It follows that $\varphi^*(y, x)$ must be stable.

(iii) Any witnesses of the failure of stability for $\varphi(x, y)$ yield witnesses of the failure of n-stability for every $n < \omega$. Thus, n-stable for some n implies stable.
On the other hand, if $\varphi(x, y)$ is n-unstable for all n, then by a straightforward compactness argument, we can see that φ is unstable.

(iv) Consider the graph G which is the disjoint union of all finite graphs. Then the edge relation E is stable in G. Indeed, if it was not, we would in particular have a vertex x_0 and infinitely many vertices $\{y_i, i \in \mathbb{N}\}$

such that $E(x_0, y_i)$ for all i. But this would imply, if G_0 is the graph containing x_0, that $y_i \in G_0$ for all i, which is impossible since G_0 is finite.

But by (iii), the edge relation is not stable in $\mathrm{Th}(G)$.

□

Definition 1.3.4. Fix $\varphi(x, y)$ in L. By a *complete φ-type over M*, $M \models T$, we mean a maximal consistent set of instances of φ and $\neg\varphi$ over M, namely L_M-formulas of the form $\varphi(x, b)$, $\neg\varphi(x, b)$ for $b \in M$. We write $S_\varphi(M)$ for the set of such complete φ-types over M.

Remark 1.3.5.

(i) By a *φ-formula over M* we mean a Boolean combination of instances (over M) of φ. For example, $(\varphi(x, c) \wedge \varphi(x, b)) \vee \neg\varphi(x, d)$ is a φ-formula over M.
(ii) Any type $p(x) \in S_\varphi(M)$ decides any φ-formula $\psi(x)$ over M, that is to say $p(x) \models \psi(x)$ or $p(x) \models \neg\psi(x)$, so in fact $p(x)$ extends to a unique maximal consistent set of φ-formulas over M.
(iii) By defining the basic open sets of $S_\varphi(M)$ to be $\{p(x) \in S_\varphi(M) : \psi(x) \in p\}$ for ψ a φ-formula, $S_\varphi(M)$ becomes a compact totally disconnected space, where in addition the clopen sets are precisely given by φ-formulas, i.e. they are the basic clopen sets.
(iv) Any $p(x) \in S_\varphi(M)$ extends to some $q(x) \in S_x(M)$ such that $p = q \upharpoonright \varphi$, where $q \upharpoonright \varphi$ is the set of φ-formulas in $q(x)$ (or instances of $\varphi, \neg\varphi$ in $q(x)$).

Definition 1.3.6.

(i) Let $p(x) \in S_x(M)$ be a complete type over M. We say that $p(x)$ is *definable* if, for each $\varphi(x, y)$ in L, there is an L_M-formula $\psi(y)$ such that for all $b \in M$, we have $M \models \psi(b)$ if and only if $\varphi(x, b) \in p$ (note that such $\psi(y)$ is unique up to equivalence). We will say that $p(x) \in S_x(M)$ is definable *over* A (for $A \subseteq M$) if each such $\psi(y)$ is over A (i.e. equivalent to a formula in L_A).
(ii) Likewise, we speak of the φ-type $p(x) \in S_\varphi(M)$ being *definable* when $\{b \in M : \varphi(x, b) \in p(x)\}$ is defined by a formula $\psi(y)$ of L_M. (Note that in this case $\psi(y)$ determines $p(x)$.) Likewise being definable over A.

As we will see later, a theory T is stable if and only if all types over *all* models of T are definable. Note that there are interesting unstable theories for which all the types over certain models are definable. For example, take T to be RCF the theory of real closed fields, and the model to be $\mathbb{R}$.

Indeed, by quantifier elimination or o-minimality, any non-realized 1-type over any model of RCF corresponds to a cut in its order. But in the case of $\mathbb{R}$, the order is complete, so for any cut, there will in fact exist a real number r such that the cut is of the form $(\{l \in \mathbb{R}, l < r\}, \{d \in \mathbb{R}, d > r\})$. Using this real number r, one can easily show definability of 1-types over $\mathbb{R}$. The higher arity case is a bit more complicated.

Another example of such a model, in the theory of p-adically closed fields, is $\mathbb{Q}_p$.

Now we come to one of the fundamental results of the subject, linking stability, definability of types, and finite satisfiability, and could be called the "fundamental theorem of stability theory". We give in the next proposition, the most general version of this result, only assuming stability of a formula in a given model. This result is a special case of a result of Grothendieck, Theorem 6 of [Gro52], after making some translations. See also [Pil16] for some more details, including reference to a paper by Ben Yaacov, who was one of the first to notice the relevance of Grothendieck's work to stability. The proof of Proposition 1.3.7 given in [Pil16] has a slight gap which is filled here. It is also amusing that Proposition 1.3.7 is closely related to and strengthens Proposition 2.2 of [Pil08] the proof of which is basically the same as Grothendieck's.

Proposition 1.3.7. *Fix a model $M \models T$ and an L-formula $\varphi(x, y)$. Then the following are equivalent:*

(i) $\varphi(x, y)$ is stable in M.
(ii) Whenever $M^* > M$ is $|M|^+$-saturated and $p(x) \in S_\varphi(M^*)$ is finitely satisfiable in M, then $p(x)$ is definable over M, moreover by some φ^*-formula over M, i.e. a Boolean combination of $\varphi(a, y)$'s, $a \in M$.

Proof. This will be a rather long proof.

We start with a discussion of some of the ingredients in the proof and make and prove a few claims. We assume to begin that $\varphi(x, y)$ is an L-formula, $M \models T$ and M^* a sufficiently saturated elementary extension of M.

Claim I. Let $p(x) \in S_\varphi(M^*)$ be finitely satisfiable in M. Then for $b \in M^*$, whether or not $\varphi(x,b) \in p$ depends on $tp_{\varphi*}(b/M)$. Hence p is determined by the function $f_p : S_{\varphi*}(M) \to \{0,1\}$ where $f_p(q) = 1$ if for some/any realization b of q in M^*, $\varphi(x,b) \in p$, and $= 0$ if for some/any realization b of q in M^*, $\neg\varphi(x,b) \in p$.

Proof of Claim I. Suppose for example $\varphi(x,b) \wedge \neg\varphi(x,c) \in p$. By finite satisfiability there is $a \in M$ such that $\models \varphi(a,b) \wedge \neg\varphi(a,c)$ whereby b,c have different φ^*-types over M. □

Claim II. Let $p(x) \in S_\varphi(M^*)$ be finitely satisfiable in M and let $f_p : S_{\varphi*}(M) \to \{0,1\}$ be as in the conclusion of Claim I. Then $p(x)$ is definable by a φ^*-formula over M iff f_p is continuous.

Proof of Claim II. This is really immediate. For example suppose f_p is continuous. Then $f_p^{-1}(1)$ is a clopen subset of $S_{\varphi*}(M)$ which has to be given by a φ^*-formula over M which will then be the required definition of p. □

The following fills the gap in [Pil16]:

Claim III. Suppose again that $p(x) \in S_\varphi(M^*)$ is finitely satisfiable in M, and now we also assume $\varphi(x,y)$ to be stable in M. Let $\psi(y)$ be a formula over M such that for all $b \in M$, if $\models \psi(b)$ then $\varphi(x,b) \in p$. Then for all $b \in M^*$ if $\models \psi(b)$, then $\varphi(x,b) \in p$.

Likewise with $\neg\varphi(x,y)$ in place of $\varphi(x,y)$.

Proof of Claim III. Suppose for a contradiction that $\psi(y)$ is over M and such that for all $b \in M$ satisfying $\psi(y)$, $\varphi(x,b) \in p$, but that there is $b' \in M^*$ such that $\models \psi(b')$ but $\neg\varphi(x,b') \in p$. Let a^* realize p.

We construct $a_1, b_1, a_2, b_2, ...$ in M such that we have that $\models \varphi(a_i, b_j)$ iff $i > j$, and such that $\models \psi(b_j)$ for all j and $\models \neg\varphi(a_i, b')$ for all i.

Suppose we have constructed $a_1, b_1, ..., a_n, b_n$ with the required properties (for $i, j = 1,..,n$). Now by assumption we have $\models \wedge_{j=1,..,n}\varphi(a^*, b_j) \wedge \neg\varphi(a^*, b')$. By finite satisfiability of p in M, there is $a_{n+1} \in M$ such that $\models \wedge_{j=1,..,n}\varphi(a_{n+1}, b_j) \wedge \neg\varphi(a_{n+1}, b')$. As M is an elementary substructure of M^* and $\models \psi(b')$ and $\wedge_{i=1,..,n}\neg\varphi(a_i, b')$ we can find $b_{n+1} \in M$ such that $\models \psi(b_{n+1}) \wedge \wedge_{i=1,..,n}\neg\varphi(a_i, b_{n+1}) \wedge \neg\varphi(a_{n+1}, b_{n+1})$. This suffices to carry out the inductive construction. □

We will now prove (ii) implies (i) of the proposition. Suppose that $\varphi(x,y)$ is not stable in M. Without loss of generality we have $a_i, b_i \in M$ for $i = 1, 2, 3, ...$ such that $\models \varphi(a_i, b_j)$ iff $i \leqslant j$. Let $\mathcal{U}$ be a nonprincipal ultrafilter on ω, and let $p^* \in S_x(M^*)$ be defined by $\chi(x) \in p^*$ iff $\{i \in \omega : \models \chi(a_i))\} \in \mathcal{U}$

for each formula $\chi(x)$ over M^*. Then p^* is a complete type over M^* which is finitely satisfiable in M. But there is no formula $\psi(y)$ over M which defines $\{b \in M^* : \varphi(x,b) \in p^*\}$. For suppose there were such a formula ψ. Then $\models \neg\psi(b_j)$ for all $j \in \omega$. By compactness there is $b \in M^*$ such that $\models \neg\psi(b) \wedge \varphi(a_i, b)$ for all $i \in \omega$. But then $\neg\varphi(x,b) \in p^*$, a contradiction. □

Finally we prove (i) implies (ii). We assume that $\varphi(x,y)$ is stable in M, M^* is a sufficiently saturated elementary extension of M, and that $p(x) \in S_\varphi(M^*)$ is finitely satisfiable in M. We want to prove that p is definable by a φ^*-formula over M. Let f_p be the function from $S_{\varphi^*}(M)$ to $\{0,1\}$ from Claim I. By Claim III, we have to prove that f_p is continuous, which is equivalent to being continuous at every $q \in S_{\varphi^*}(M)$. So fix $q(y) \in S_{\varphi^*}(M)$. Without loss of generality $\varphi(x,b) \in p$ for b realizing q in M^*. We have to find a formula $\psi(y) \in q(y)$ such that for all $b' \in M^*$, $\varphi(x,b') \in p(x)$. By Claim III, it suffices to find a formula $\psi(y) \in q$ such that for every $b' \in M$ satisfying $\psi(y)$, $\varphi(x,b') \in p(x)$. We will assume this does not hold, and get a contradiction to stability of $\varphi(x,y)$ in M.

So the current situation is that $q(y) \in S_{\varphi^*}(M)$, and for b^* say realizing q in M^*, $\varphi(x,b^*) \in p(x)$, but for every formula $\psi(y) \in q(y)$, there is $b \in M$ satisfying q such that $\neg\varphi(x,b) \in p(x)$. Let a^* realize p (in a bigger model than M^*). We will construct inductively (similar to the proof of Claim III), $a_i, b_i \in M$ for $i = 1,2,3,\ldots$ such that

(1) $\models \varphi(a_i, b_j)$ iff $i \leqslant j$,

(2) $\models \neg\varphi(a^*, b_j)$ for all j, and

(3) $\models \varphi(a_i, b^*)$ for all i.

Suppose we have already produced $a_1, b_1, \ldots, a_n, b_n$ satisfying (1), (2), (3) for $i,j = 1,\ldots,n$. So we have $\models \varphi(a^*,b^*) \wedge \wedge_{j=1,\ldots,n}\neg\varphi(a^*,b_j)$. By finitely satisfiability of p in M, let $a_{n+1} \in M$ be such that $\models \varphi(a_{n+1}, b^*) \wedge \wedge_{j=1,\ldots,n}\neg\varphi(a_{n+1}, b_j)$. Now, as we also have $\models \varphi(a_i, b^*)$ for $i = 1,\ldots,n$, choose $b_{n+1} \in M$ such that $\models \varphi(a_i, b_{n+1})$ for $i = 1,\ldots,n+1$. This completes the induction.

We have a contradiction to $\varphi(x,y)$ being stable in M. So we have completed the proof of (i) implies (ii) and so of the proposition. □

Remark 1.3.8. We make a few remarks:

(1) The proof of 1.3.7 above shows that the definition of $p(x)$ (over M) can be chosen to be a *positive* φ^*-formula over M (a positive Boolean combination of instances of φ^* over M).

(2) The proof of 1.3.7 (i) implies (ii) could be rearranged to first give definability (by a φ^*-formula) of any $p(x) \in S_\varphi(M)$, and then deduce (ii) from Claim III.
(3) The version of 1.3.7 for complete types: *if every formula is stable in M, then any complete type over M^* finitely satisfiable in M is definable over M* is a consequence of Proposition 2.2 and Corollary 2.3 in [Pil08], which has a more indirect proof.
(4) 1.3.7 is basically enough to develop independence and (non) forking in stable theories when the base is a model. But this will ignore the whole machinery of strong types.

We now consider some consequences of 1.3.7 when $\varphi(x; y)$ is stable for T.

Proposition 1.3.9. *Let $\varphi(x; y)$ be an L-formula. Then the following are equivalent:*

1. *$\varphi(x; y)$ is stable (for T).*
2. *For all $M \models T$ and $p(x) \in S_\varphi(M)$, $p(x)$ is definable, i.e. there is some L_M-formula $\psi(y)$ such that for all $b \in M$, $M \models \psi(b)$ if and only if $\varphi(x, b) \in p(x)$.*
3. *For all cardinal $\lambda \geqslant |T|$ and $M \models T$ of cardinality λ, we have $|S_\varphi(M)| \leqslant \lambda$.*
4. *There is $\lambda \geqslant |T|$ such that for all $M \models T$ of cardinality λ, we have $|S_\varphi(M)| \leqslant \lambda$.*

Proof. 1 $\Rightarrow$ 2: This is a consequence of Proposition 1.3.7. Indeed, fix $p(x) \in S_\varphi(M)$, then p has an extension to $p' \in S_\varphi(M^*)$ which is finitely satisfiable in M (where M^* is sufficiently saturated). By Proposition 1.3.7, we obtain that p' is definable over M, so $p = p'|M$ is also definable.

2 $\Rightarrow$ 3: Each $p(x) \in S_\varphi(M)$ is determined by its definition $\psi_p(y) \in L_M$. But there are at most λ choices for such a definition, hence the result.

3 $\Rightarrow$ 4: Immediate.

4 $\Rightarrow$ 1: We prove the contrapositive. Assume $\varphi(x, y)$ is unstable. Let $\lambda \geqslant |T|$. We can find a total ordering $(I, <)$ of cardinality λ with $> \lambda$ initial segments. (For example, let $\mu \leqslant \lambda$ be the least cardinal such that $2^\mu > \lambda$. Let I be the set of eventually constant functions from μ to 2, with the lexicographic ordering.) By compactness (and instability of φ) we can find a_J, b_i in $\bar{M}$ for $i \in I$ and J an initial segment of I, such that $\models \varphi(a_J, b_i)$ iff $i \in J$. Let M be a model (elementary substructure of $\bar{M}$) containing all the b_i and of cardinality λ (by downward Löwenheim-Skolem). Then the $tp_\varphi(a_J/M)$ give $> \lambda$ complete φ-types over M. □

For the next results, we will need a new tool: Cantor–Bendixson rank, or CB rank. Let us define it:

Definition 1.3.10 (Cantor–Bendixson Rank). Let X be a topological space. The Cantor–Bendixson rank is a function $CB_X : X \to \mathrm{On} \cup \{\infty\}$ (where On is the class of ordinals). Let $p \in X$, then:

(*i*) $CB_X(p) \geqslant 0$,
(*ii*) $CB_X(p) = \alpha$ if $CB_X(p) \geqslant \alpha$ and p is isolated in the (closed) subspace $\{q \in X : CB_X(q) \geqslant \alpha\}$.
(*iii*) $CB_X(p) = \infty$ if $CB_X(p) > \alpha$ for every ordinal α.

For example, $CB_X(p) = 0$ if p is isolated, equivalently if $\{p\}$ is open. $CB_X(p) \geqslant 1$ otherwise.

Note that in (*ii*) we say that the subspace $\{q \in X : CB_X(q) \geqslant \alpha\}$ is closed for all α. This is a consequence of the fact that the set of isolated points of any topological space form an open set, as a union of open sets.

Fact 1.3.11. *Suppose X is compact and $CB_X(p) < \infty$ for every p in X. Then there exists a maximal element α of $\{CB_X(p) : p \in X\}$ and $\{p \in X : CB_X(p) = \alpha\}$ is finite and non empty. We then define $CB(X) = \alpha$.*

Proof. Assume there is no maximal element. Then, for each ordinal α, there exists some p_α in X such that $CB_X(p_\alpha) > \alpha$. The set $\{p_\alpha, \alpha \in \mathrm{On}\}$ must have an accumulation point p in the compact set X, which cannot be isolated in any of the $\{q \in X : CB_X(q) \geqslant \alpha\}$. Hence $CB_X(p) = \infty$, a contradiction.

Let $\alpha = \sup\{CB_X(p) : p \in X\}$. We want to show that $X_\alpha = \{p \in X : CB_X(p) = \alpha\}$ is non-empty. Suppose not. Then α has to be a limit ordinal. For each β less than α, we consider $X_{<\beta} = \{p \in X : CB_X(p) < \beta\}$. Now, the collection $\mathcal{C} = \{X_\beta : \beta < \alpha\}$ is an open cover of X which clearly has no finite subcover, as α is a limit ordinal. This contradicts the assumption that X is compact. We have shown that X_α is non-empty.

The subset $\{p \in X : CB_X(p) \geqslant \alpha\}$ is closed, so compact. Since α is maximal, all points in $\{p \in X : CB_X(p) \geqslant \alpha\}$ are isolated. Therefore, $\{p \in X : CB_X(p) \geqslant \alpha\}$ is finite. □

Lemma 1.3.12. *Suppose $\varphi(x, y)$ is stable for T. Let $M \models T$. Let $X = S_\varphi(M)$. Then, $CB_X(p) < \infty$ for each p in X. (So $CB(S_\varphi(M)) < \infty$.)*

Proof. Define $X_\alpha = \{p \in X : CB_X(p) \geqslant \alpha\}$. Assume that there exists some q such that $CB_X(q) = \infty$, then for some α, $X_\alpha \neq \emptyset$ and has no isolated points.

We now fix such an α. Since there are no isolated points in X_α, we can find $p_0, p_1 \in X_\alpha$ where $p_0 \neq p_1$. Since $S_\varphi(M)$ is Hausdorff, we can find $\psi_0(x)$ such that $\psi_0(x) \in p_0$ and $\psi_1(x) = \neg\psi_0(x) \in p_1$. Then $\psi_0(x)$, $\psi_1(x)$ both define nonempty clopen subsets of X_α. Again as the latter has no isolated points, we can find φ-formulas over M, $\psi_{00}(x)$, $\psi_{01}(x)$, $\psi_{10}(x)$, $\psi_{11}(x)$, which are pairwise inconsistent, each of which defines a nonempty clopen subset of X_α and with each of $\psi_{00}(x)$ and $\psi_{01}(x)$ implying $\psi_0(x)$, and each of $\psi_{10}(x)$, $\psi_{11}(x)$ implying $\psi_1(x)$. Continuing this way we build a suitable tree of φ-formulas over M, $\psi_\eta(x)$ for $\eta \in 2^{<\omega}$, each branch of which is consistent extends to a complete type in $S_\varphi(M)$. This produces continuum many complete φ-types over M. Working in a countable sublanguage L_0 of L containing φ and choosing a countable L_0-elementary substructure M_0 of M such that all the ψ_η have parameters from M_0, we obtain a countable theory T_0 with continuum many complete φ-types over some countable model. This contradicts Proposition 1.3.9, as φ is also stable for T_0. □

Remark 1.3.13. A similar proof shows that if $\varphi(x; y)$ is stable in M, then every p in $S_\varphi(M)$ has CB-rank, using 1.3.7.

We have seen that if $\varphi(x; y)$ is stable, $p(x) \in S_\varphi(M)$, and $M \models T$, then $p(x)$ is definable. We now introduce the obvious notion of being definable over a subset A of M.

Definition 1.3.14. If $A \subset M$ we say that $p(x) \in S_\varphi(M)$ is definable over A if p is definable and a defining formula $\psi(y)$ can be found in L_A (i.e. "over A"). Likewise for a complete type $p(x) \in S(M)$ to be definable over A.

Lemma 1.3.15. *Assume $\varphi(x, y)$ is stable. Let $M \models T$, $A \subseteq M$ and let $p(x) \in S_x(A)$. Then there is $q(x) \in S_\varphi(M)$ such that $p(x) \cup q(x)$ is consistent and $q(x)$ is definable "almost over A", i.e. over $\mathrm{acl}^{eq}(A)$.*

Proof. We may assume that M is $||T| + |A||^+$-saturated and strongly homogeneous, since every model of T is contained in such a model.

Consider the set $X = \{q(x) \in S_\varphi(M) : p(x) \cup q(x) \text{ is consistent}\}$. We leave it as an exercise to show that X is a closed subset of $S_\varphi(M)$. Using 1.3.12 one shows that $CB(X) < \infty$. Let, by Fact 1.3.11, $X_0 \subset X$ be the finite set of elements of maximum CB_X-rank α. Let $q(x) \in X_0$ and let $\psi(y)$ be its definition (i.e. $M \models \psi(b)$ if and only if $\varphi(x, b) \in q$), which is a priori over M. Note that $\mathrm{Aut}(M/A)$ acts on X, and X_0 is invariant under this action. Hence $q(x)$ has only finitely many images under $\mathrm{Aut}(M/A)$. So $\psi(y)$ has finitely many images under $\mathrm{Aut}(M/A)$, and Lemmas 1.2.13 and 1.2.35 yield that $\psi(y)$ is almost over A. □

Remark 1.3.16. When $A = M_0 \prec M$, the previous lemma is immediate from 1.3.7.

Theorem 1.3.17 (Symmetry). *Let $\varphi(x, y)$ be stable, $p(x) \in S_\varphi(M)$ and $q(y) \in S_{\varphi^*}(M)$. Let $\psi(y)$ be the φ^*-formula defining $p(x)$ and $\chi(x)$ the φ-formula defining $q(y)$. Then $\psi(y) \in q(y) \Leftrightarrow \chi(x) \in p(x)$.*

Proof. Note first that by 1.3.7, ψ, χ exist and are φ^*, φ formulas, respectively. We will show that $\psi \in q$ implies $\chi \in p$, the other implication being proved exactly the same way.

Let $M \prec M^*$ where M^* is $|M|^+$-saturated. Suppose $\psi(y) \in q$. Let b realize q in M^*. Let $p^* \in S_\varphi(M^*)$ be the extension of p given by the same defining formula $\psi(y)$. By 1.3.7, p^* is finitely satisfiable in M and of course $\varphi(x, b) \in p^*$.

Suppose for a contradiction that $\neg\chi(x) \in p$. So $\neg\chi(x) \in p^*$ and so $\neg\chi(x) \wedge \varphi(x, b) \in p^*$. By finite satisfiability, there is $a \in M$ such that $\models \neg\chi(a) \wedge \varphi(a, b)$. But then $\varphi(a, y) \in q(y)$ and $M \models \neg\chi(a)$, a contradiction to the fact that $\chi(x)$ is the defining formula for $q(y)$. □

Remark 1.3.18. The proof shows that the above theorem holds only under the assumption of stability of $\varphi(x, y)$ in M.

Definition 1.3.19.

1. Let $A \subseteq \bar{M}$ (or $\bar{M}^{eq}$), and let $\varphi(x, y)$ be a formula. By a φ-formula over A, we mean a φ-formula which is equivalent to a formula with parameters in A. (Note that this is consistent with our earlier definition when A is a model M, because if some Boolean combination of instances of φ is equivalent to a formula with parameters from M then it is equivalent to a Boolean combination of instances of φ where the parameters appearing in these instances, are from M.)
2. A complete φ-type over A is a maximal consistent collection of φ-formulas over A.
3. Given a complete φ-type over B and $A \subseteq B$, $p{\restriction}A$ is the complete φ-type over A consisting of formulas in p which are over A.

Lemma 1.3.20 (Uniqueness). *Let $\varphi(x, y)$ be stable in M and let $A = acl^{eq} \subseteq M$. If p_1, p_2 are complete φ-types over M, definable over A and $p_1 \restriction A = p_2 \restriction A$, then $p_1 = p_2$.*

Proof. We show that for all $b \in M$, $\varphi(x, b) \in p_1 \Leftrightarrow \varphi(x, b) \in p_2$. Let $\psi_i(y)$ be the φ-definition of p_i, a φ^*-formula over A. Fix an arbitrary $b \in M$, let $q(y) = \mathrm{tp}_{\varphi^*}(b/A)$. By Lemma 1.3.15, there is some $q' \in S_{\varphi^*}(M)$ extending q and definable over A.

Let $\chi(x)$ be the defining formula for q', which will be a φ-formula over A. Then by 1.3.17, for each $i = 1, 2$, $\chi(x) \in p_i(x)$ iff $\psi_i(y) \in q'(y)$. But as p_1 and p_2 have the same restriction to A and $\chi(x)$ is a φ-formula over A, we have that $\chi(x) \in p_1$ iff $\chi(x) \in p_2$. Hence $\psi_1(y) \in q'$ iff $\psi_2(y) \in q'$. As $\psi_i(y)$ are φ^*-formulas over A and $q = tp_{\varphi^*}(b/A)$ it follows that $\models \psi_1(b)$ iff $\models \psi_2(b)$, so $\varphi(x, b) \in p_1$ iff $\varphi(x, b) \in p_2$. □

Remark. Together with existence, uniqueness, this shows that if A is an algebraically closed subset of M and $p \in S_\varphi(A)$ then p has a unique extension $p'(x) \in S_\varphi(M)$ which is definable over A.

In the following theorem, by a finite equivalence relation we mean one with finitely many classes.

Theorem 1.3.21 (Conjugacy and the Finite Equivalence Relation Theorem). *Suppose that $A \subseteq M$, $p \in S_\varphi(A)$, and X is the set of all complete φ-types over M extending p which are definable over $acl^{eq}(A)$.*

1. *X is finite*
2. $\mathrm{Aut}(M/A)$, *the group of automorphisms of M which fix A pointwise, acts transitively on X if M is sufficiently homogeneous and saturated.*
3. *There exists an A-definable finite equivalence relation E, each of whose classes is defined by a φ-formula, such that for any $q_1, q_2 \in X$, we have $q_1 = q_2$ if and only if $q_1(x_1) \cup q_2(x_2) \models E(x_1, x_2)$.*

Proof. Note first that by uniqueness above, X is in natural bijective correspondence with $Y := \{q \restriction \mathrm{acl}^{eq}(A) : q \in X\}$. Moreover, Y is still acted upon by $\mathrm{Aut}(M/A)$.

Proof of 2. It suffices to show that $\mathrm{Aut}(M/A)$ acts transitively on Y. Let us fix a complete type $p'(x)$ over A extending (containing) p.

Claim. *For each $q(x) \in Y$, $p'(x) \cup q(x)$ is consistent.*

Proof of claim. If not there is a formula $\psi(x) \in q$ such that $p'(x)$ implies $\neg\psi$. So $p'(x)$ implies each of the finitely many $Aut(M/A)$-conjugates of $\neg\psi$. The disjunction of these finitely many conjugates is then a φ-formula over A implied by p', hence in p, a contradiction. The claim is established.

Now let $q_1, q_2 \in Y$. By the claim, there are realizations b_1, b_2 of q_1, q_2 in M, such both b_1 and b_2 have the same complete type p' over A. But then, by our assumptions on M, there is an automorphism α of M, fixing A pointwise and taking b_1 to b_2. But then $\alpha(q_1) = q_2$. This completes the proof of (ii).

Proof of 1. It suffices to show that Y is finite. Let $q \in Y$, it is $\text{acl}^{eq}(A)$-definable, and so has only finitely many images under $\text{Aut}(M/A)$. Since the action of $\text{Aut}(M/A)$ is transitive, Y is the orbit of q, which is finite.

Proof of 3. Each two types $q_i, q_j \in Y$ are separated by some formula $\theta_{i,j}(x)$, i.e. $\theta_{i,j}(x) \in q_i$ and $\neg\theta_{i,j}(x) \in q_j$. Consider the set of formulas $\{\theta_{i,j} | q_i, q_j \in Y, i \neq j\}$, it is finite. We can close this set under $\text{Aut}(M/A)$, to obtain a new set of formulas, which we denote Θ, which is also finite.

Let $E(x_1, x_2)$ be the equivalence relation defined by $\bigwedge_{\theta \in \Theta} \theta(x_1) \leftrightarrow \theta(x_2)$. Then E is A definable (as it is invariant under automorphisms of the monster model fixing A pointwise), and obviously has finitely many classes, each of which is defined by a φ-formula (over $acl^{eq}(A)$). Now any q in X determines an E-class, but by construction and the fact that the q is determined by its restriction to $acl^{eq}(A)$, in turn q is determined by its E-class. □

We will now introduce the notions of forking and dividing.

Definition 1.3.22. Let $A \subset \bar{M}$, and let $\varphi(x, b)$ be a formula where we witness the parameter $b \in \bar{M}$.

(i) We say that $\varphi(x, b)$ divides over A if there is an A-indiscernible sequence $(b_i)_{i \in \mathbb{N}}$ such that $b_0 = b$ and the set $\{\varphi(x, b_i) : i \in \omega\}$ is inconsistent. By compactness and indiscernibility, this is the same as saying the above set is k-inconsistent for some k.

(ii) We say that $\varphi(x, b)$ forks over A if there are $\psi_1(x, b_1), ..., \psi_k(x, b_k)$ each dividing over A and $\models \varphi(x, b) \to \bigvee_i \psi_i(x, b_i)$.

Note that if $b \in A$ and $\varphi(x, b)$ is consistent, then $\varphi(x, b)$ does not divide over A, as any A-indiscernible sequence containing b will just repeat b.

Remark 1.3.23. We can extend the above definition to partial types over B. That is, if $\pi(x,b)$ is a partial type over B where we witness by the (possibly infinite) tuple b the parameters in π, then $\pi(x,b)$ divides over A iff there is an A-indiscernible sequence $(b_i : i < \omega)$ of realizations of $tp(b/A)$ such that $\cup_i\{\pi(x,b_i) : i < \omega\}$ is inconsistent. Then, assuming $\pi(x,b)$ to be closed under conjunctions, by compactness $\pi(x,b)$ divides over A iff some formula $\varphi(x,b) \in \pi(x,b)$ divides over A. We will say that a partial type forks over A if it implies a finite disjunction of formulas each of which divides over A. Again by compactness this is equivalent to a single formula in $\pi(x)$ forking over A.

Let $\pi(x)$ be a partial type over B, which we assume, without loss of generality, to be closed under conjunctions. Then $\pi(x)$ divides/forks over A if there is $\varphi(x) \in \pi(x)$ that divides/forks over A.

Also the notion of dividing over A depends only on the definable set defined by $\varphi(x,b)$ rather than the specific shape of the formula.

One idea behind this notion, is that if $\varphi(x,b)$ divides over A then $\varphi(x,b)$ is "very far" from being over A.

The following exercises give the meaning in special cases.

Example. Let k be a countable subfield of $\mathbb{C}$. Let V be a complex variety, a definable set in the complex field. Then V does not divide over k if and only if V does not fork over k if and only if V is defined over $\operatorname{acl}(k)$.

Example. Let T be the theory of an equivalence relation with infinitely many infinite classes. Then for every b, the formula $E(x,b)$ divides over $\varnothing$.

The following lemma will be useful in proofs:

Lemma 1.3.24. *Let $(b_i)_{i\in\mathbb{N}}$ be an A-indiscernible sequence. Then for all i, we have* $\operatorname{tp}(b_i/\operatorname{acl}^{eq}(A)) = \operatorname{tp}(b_0/\operatorname{acl}^{eq}(A))$.

Proof. Suppose not. Then there exists an A-definable finite equivalence relation $E(y,z)$ and some $i > 0$ such that $\neg E(b_0,b_i)$. By indiscernibility, this implies that $\neg E(b_j,b_k)$ for all $j,k \in \mathbb{N}$, which contradicts E being finite. □

Corollary 1.3.25.

(i) If $\varphi(x,b)$ is over $\operatorname{acl}^{eq}(A)$, *then it does not divide over A.*
(ii) $\varphi(x,b)$ divides over A iff $\varphi(x,b)$ divides over $acl^{eq}(A)$.

The following is obvious but is included for the record.

Lemma 1.3.26. *Let $p \in S_\varphi(M)$ be a definable type, and $M \prec M'$ an elementary extension of M. There is a unique type $p' \in S_\varphi(M')$ with same definition as p. Moreover p' extends p.*

Proof. Let $\psi(y)$ be the defining formula of p. Let $p'(x)$ be defined by $\{\varphi(x,b) : b \in M', \models \psi(b)\} \cup \{\neg\varphi(x,b) : b \in M', \models \neg\psi(b)\}$.

We only have to show consistency of p', so by compactness, finite consistency. But if some finite subset of p' is inconsistent, then as the formula ψ is over M, and $M \prec M'$, we will contradict the consistency of p. □

We now come to the rather delicate issue of the relation between the theory we have obtained so far (definability, finite satisfiability), and the notions of forking and dividing. We will give a couple of closely related results, with amusing proofs some of which introduce new notions.

Proposition 1.3.27. *Suppose $\varphi(x,y)$ is stable, and $A \subseteq M$ where M is κ-saturated and κ-strongly homogeneous for some $\kappa > 2^{(|T|+|A|)}$. Let $p(x) \in S_\varphi(M)$. The following are equivalent:*

(1) p is definable over $acl^{eq}(A)$.
(2) p does not divide over A.

Proof. (1) implies (2): Pick any formula $\chi(x,c)$ in p, where we witness the parameters c. Let $(c = c_1, c_2, ...)$ be an A-indiscernible sequence which we may assume (by saturation) to be in M. As p is definable over $acl^{eq}(A)$, and by Lemma 1.3.24, all c_i have the same type over $acl^{eq}(A)$ it follows that each $\chi(x,c_i) \in p$ and hence $\{\chi(x,c_i) : i = 1,2,...\}$ is consistent.

(2) implies (1): We will make use of a certain generalization of strong types which are called Lascar strong types. One can see [TZ12] for additional details and background, including references to the original papers. Anyway given tuples b, c of the same length (finite or ω) we say that b and c have the same Lascar strong type over A, if there are tuples (of the same length), $b_0, b_1, \ldots, b_m$ for some finite m, such that $b_0 = b$, $b_m = c$, and for $i = 1, \ldots, m-1$, b_i, b_{i+1} are the first two elements of an infinite, A-indiscernible sequence. $Aut(M/A)$ acts on the collection of Lascar strong types and $Autf(M/A)$ is by definition the subgroup of $Aut(M/A)$ which fixes all the Lascar strong types. It is a normal subgroup of $Aut(M/A)$ and the quotient is denoted $Gal_L(T_A)$. The main point here is that $Gal_L(T_A)$ has cardinality at most $2^{(|T|+|A|)}$, in particular $< \kappa$. Equivalently the index of $Autf(M/A)$ in $Aut(M/A)$ is at most $2^{(|T|+|A|)}$.

Now assume that $p(x) \in S_\varphi(M)$ does not divide over A.

Claim. *Suppose $b_1, b_2, b_3, \ldots$ are tuples in M of the same length as the tuple y of variables, and that $(b_1, b_2, \ldots)$ is A-indiscernible. Then $\varphi(x, b_1) \in p(x)$ if and only if $\varphi(x, b_2) \in p(x)$.*

Proof of Claim. Suppose otherwise. So (without loss) $\varphi(x, b_1) \wedge \neg\varphi(x, b_2) \in p(x)$. But then, as $((b_1, b_2), (b_3, b_4), (b_5, b_6), \ldots)$ is also A-indiscernible, it follows, from p not dividing over A, that $\{\varphi(x, b_i) \wedge \neg\varphi(x, b_{i+1}) : i = 1, 3, 5, \ldots\}$ is consistent. It is an exercise (using indiscernibility) to show that this contradicts the stability of $\varphi(x, y)$. The Claim is proved.

It follows that whenever b_1 and b_2 have the same Lascar strong type over A then $\varphi(x, b_1) \in p$ iff $\varphi(x, b_2) \in p$. This means that p is invariant under $Autf(M/A)$. Let $\psi(y)$ be the definition of p (i.e. for any $b \in M$, $\varphi(x, b) \in p$ iff $M \models \psi(b)$). So ψ is fixed by $Autf(M/A)$. So as the index of $Autf(M/A)$ in $Aut(M/A)$ is of cardinality $< \kappa$, $\psi(y)$ has $< \kappa$ many images under $Aut(M)$, hence by 1.2.13 is over $acl^{eq}(A)$. □

Here is a related result about formulas rather than complete φ-types.

Lemma 1.3.28. *Let $\varphi(x, y)$ be stable, and $b \in \bar{M}$. The following are equivalent:*

(1) There is some M containing $A \cup \{b\}$ and $p(x) \in S_\varphi(M)$ such that $p(x)$ is defined almost over A and $\varphi(x, b) \in p$.
(2) $\varphi(x, b)$ does not divide over A.

Proof. (1) implies (2) follows by a similar proof to that of (1) implies (2) in the previous proposition.

(2) implies (1): We assume (2). We may assume $A = acl^{eq}(A)$. Let M be a model (elementary substructure of $\bar{M}$) containing $A \cup \{b\}$. By existence Lemma 1.3.15, let $q(y) \in S_{\varphi^*}(M)$ be definable over A and consistent with $tp(b/A)$. Let $\bar{q}(y)$ be a complete type over M extending $q(y) \cup tp(b/A)$. Finally let M^* be an $|M|^+$-saturated model containing M and $q^*(y)$ a complete type over M^* extending q and finitely satisfiable in M. Let $b_1, b_2, b_3, \ldots$ in M^* be a "coheir sequence" for q^*, namely b_1 realizes $q^* \upharpoonright M = \bar{q}$ and b_{i+1} realizes $q^* \upharpoonright (M, b_1, \ldots, b_i)$. Note that $tp(b_i/A) = tp(b/A)$ for all i. By Proposition 1.2.9, $(b_i : i = 1, 2, \ldots)$ is M-indiscernible, so also an A-indiscernible sequence of realizations of $tp(b/A)$.

Let us note that, by 1.3.7, $q^* \upharpoonright \varphi^*$ (the restriction of q^* to a complete φ^*-type over M^*) is definable over M (by a φ-formula). However the restriction of $q^* \upharpoonright \varphi^*$ to M is precisely q which is definable over A. Hence

$q^* \upharpoonright \varphi^*$ is definable over A, and let $\chi(x)$ be the φ-formula over A which is the definition.

By our assumption that $\varphi(x, b)$ does not divide over A, $\{\varphi(x, b_i) : i = 1, 2, ...\}$ is consistent, so let $a \in M^*$ realize it. Let $p(x) \in S_\varphi(A)$ be the complete φ-type of a over A. Let (by existence), $p^* \in S_\varphi(M^*)$ extend p and be definable over A. Let $\psi(y)$ be the φ^*-formula over A defining p^*.

It is now enough to prove:

Claim. $\varphi(x, b) \in p^*(x)$.

Proof of Claim. Suppose, not. Then $\models \neg\psi(b)$. So (as $q^* \upharpoonright \varphi^*$ extends $tp_{\varphi^*}(b/A)$), $\neg\psi(y) \in q^* \upharpoonright \varphi^*$. By 1.3.17, $\neg\chi(x) \in p^*$, which implies that $\models \neg\chi(a)$ and so $\neg\varphi(a, y) \in q^* \upharpoonright \varphi^*$, in particular $\neg\varphi(a, y) \in q^*$. Now let M^{**} be a an $|M^*|^+$-saturated elementary extension of M^*. Let q^{**} be the *unique* extension of q^* to M^{**} which is finitely satisfiable in M. (The reader can check the uniqueness claim here.) Now, let $b_\omega, b_{\omega+1}, ...$ be a "coheir sequence" of q^{**} over M^* (that is, as usual $b_{\omega+(i+1)}$ realizes the restriction of q^{**} to $M^*, b_\omega, ..., b_{\omega+i}$). Then $(b_1, b_2, ..., b_\omega, b_{\omega+1}, ...)$ is a coheir sequence of q^{**} over M, so is indiscernible over M (again by Proposition 1.2.9). However, as each $b_{\omega+i}$ realizes $q^*|\varphi^*$, and $\models \neg\chi(a)$ we have $\models \neg\varphi(a, b_{\omega+i})$ for all i. On the other hand by choice of a we have $\models \varphi(a, b_i)$ for all of i. This, as well as the indiscernibility of $(b_1, b_2, ..., b_\omega, b_{\omega+1}, ...)$ contradicts stability of $\varphi(x, y)$, as the reader is invited to verify. This completes the proof of the claim and the theorem. □

The same method of proof gives the following which is close to 1.3.27 above.

Lemma 1.3.29. *Assume $\varphi(x, y)$ is stable, fix A small and $p \in S_\varphi(A)$. Let $M \supset A$ and $b \in M$. The following are equivalent:*

(i) $\varphi(x, b) \in p^*(x)$ for some $p^*(x) \in S_\varphi(M)$ extending p and definable almost over A.

(ii) $p(x) \cup \{\varphi(x, b)\}$ does not divide over A.

Let us remark that (trivially) dividing implies forking, but the converse is not true in general.

Example 1.3.30. Consider the unit circle $\mathbb{S}^1$, together with the ternary relation B of betweeness: we say that $B(x, y, z)$ is the counterclockwise arc from x to z goes through y.

Then the formula $x = x$ does not divide over the empty set, as it is over the empty set. However, it does fork over the empty set. Indeed, it implies, for any $a, b \in \mathbb{S}^1$, the disjunction $B(b, x, a) \vee B(a, x, b)$, and each of these two formulas is easily verified to divide over the empty set.

However, the equivalence between forking and dividing will be true for stable formulas in a certain strong sense.

We start with the following which is part of the rationale for the definition of forking.

Remark 1.3.31. Suppose the partial type $\Phi(x, c)$ does not fork over A (where c is a possibly infinite tuple). Then for any model M containing $A \cup \{c\}$ there is a complete type $p(x)$ over N which extends $\Phi(x, c)$ and does not divide over A.

Proof. Consider the set of formulas $\Sigma(x) = \Phi(x, c) \cup \{\neg\varphi(x), \varphi \text{ over } M, \varphi \text{ divides over } A\}$. It is consistent. For if not, by compactness we would have $\Phi(x, c) \models \varphi_1(x) \vee \cdots \vee \varphi_n(x)$, for some φ_i over M, dividing over A, contradicting $\Phi(x, c)$ not forking over A.

Therefore, we can extend Σ to a complete type over N, which is the type we were looking for. □

We can now state and prove:

Theorem 1.3.32. *Suppose $\varphi(x, y)$ is stable. Then for any b, the formula $\varphi(x, b)$ divides over A if and only if it forks over A in the strong sense that it implies a disjunction of φ-formulas each of which divides over A.*

Proof. The left to right direction is immediate. For the other direction, suppose that $\varphi(x, b)$ does not divide over A. Let M^* be a $(|T| + |A|)^+$-saturated model containing $A \cup \{b\}$. By Lemma 1.3.28, there is $p(x) \in S_\varphi(M^*)$ definable almost over A, containing $\varphi(x, b)$. In particular it is invariant under $\mathrm{Aut}(M^*/\mathrm{acl}^{eq}(A))$, because any $\sigma \in \mathrm{Aut}(M^*/\mathrm{acl}^{eq}(A))$ will fix the definition of p.

By way of contradiction, suppose that $\varphi(x, b) \models \psi_1(x, b_1) \vee \ldots \vee \psi_r(x, b_r)$, with each ψ_i a φ-formula dividing over A. Note that if $\mathrm{tp}(b_i/Ab) = \mathrm{tp}(b_i'/Ab)$ for some b_i', then $\psi(x, b_i')$ divides over A as well. Hence, we can assume that $b_i \in M^*$ for all i, by saturation of M^*.

As $\varphi(x, b) \in p$, we have $\psi_i(x, b_i) \in p$ for some i. But by (the easy direction of) Proposition 1.3.27, this implies that $\psi_i(x, b_i)$ does not divide over A, a contradiction. □

Putting everything together we obtain:

Corollary 1.3.33. *Suppose $\varphi(x, y)$ is stable, $A \subset M$, and $p(x) \in S_\varphi(M)$. Then the following are equivalent:*

(i) p does not fork over A,
(ii) p does not divide over A,
(iii) p is definable almost over A.

Proof. By 1.3.32 (and the fact that any φ-formula $\psi(x)$ is of the form $\chi(x, c)$ where $\chi(x, z)$ is a stable L-formula) (i) and (ii) are equivalent. Also (iii) implies (ii) by the usual easy argument (or by 1.3.28).

Now assume (i). By Remark 1.3.31, p extends to a complete φ-type p' over a sufficiently saturated model M' containing M, which does not divide over A. By 1.3.27, p' is definable almost over A, hence so is its restriction p. □

We now return to ranks, specifically the Cantor–Bendixson rank. Recall that this rank was defined in 1.3.10, and some of its elementary properties were proven. For stable formulas, the Cantor–Bendixson rank is finite:

Lemma 1.3.34. *Suppose $\varphi(x, y)$ is stable, and let $X = S_\varphi(\overline{M})$. Then* CB$(X)$ *is finite.*

Proof. Suppose for contradiction that there is $p \in S_\varphi(\overline{M})$ with $\mathrm{CB}(p) \geqslant \omega$. Fix n, so $CB_X(p) \geqslant n$. As in Lemma 1.3.12, we can build, a tree of formulas $\{\varphi_\mu(x), \mu \in^{n\geqslant} 2\}$ such that $\varphi_\varnothing(x)$ is $x = x$, for $0 \leqslant l(\mu) < n$ the formulas $\varphi_{\mu\wedge 0}$ and $\varphi_{\mu\wedge 1}$ are of the form $\varphi(x, b), \neg\varphi(x, b)$ or $\neg\varphi(x, b), \varphi(x, b)$ and moreover, for each $\mu \in^n 2$, the set of formulas $\{\varphi_{\mu|_i}(x), i \leqslant n\}$ is consistent.

As we can do this for every n, by compactness and saturation, we find a tree $\{\varphi_\mu, \mu \in^{\omega\geqslant} 2\}$ such that $\varphi_\varnothing(x)$ is $x = x$, and for each η, the pair $\varphi_{\eta\wedge 0}, \varphi_{\eta\wedge 1}$ is of the form $\varphi(x, b), \neg\varphi(x, b)$ or $\neg\varphi(x, b), \varphi(x, b)$ for some b, and such that moreover we have consistency of each branch $\{\varphi_{\eta|n}(x) : n < \omega\}$ for each $\eta \in^\omega 2$. We get a contradiction to stability as at the end of the proof of Lemma 1.3.12. □

The Cantor–Bendixson rank can be extended to define a rank for partial types:

Definition 1.3.35. Fix $\varphi(x, y)$ a stable formula. Let Φ be a partial φ-type over some small set of parameters. We define $R_\varphi(\Phi) = \max\{\mathrm{CB}_X(p), p \in S_\varphi(\overline{M}), \Phi \subset p\}$, where $X = S_\varphi(\overline{M})$.

A justification for the definition is that Φ defines a closed so compact subset of X, hence the supremum is attained witnessed by finitely many types.

Proposition 1.3.36. *Let $\varphi(x, y)$ be stable, $p \in S_\varphi(A)$, with $A \subset M$. Let $p'(x)$ be an extension of $p(x)$ over M. Then $p'(x)$ is definable almost over A if and only if $R_\varphi(p) = R_\varphi(p')$.*

To prove this, we will need one more fact about local stability theory.

Lemma 1.3.37. *Let $\varphi(x, y)$ be stable, and $\varphi(x, b)$ be some instance of it. Then $\varphi(x, b)$ does not divide over A if and only if some finite positive Boolean combination of* $\mathrm{Aut}(\overline{M}/A)$ *conjugates of $\varphi(x, b)$ is consistent and over A.*

Proof. Suppose $\varphi(x, b)$ does not divide over A. Let $M^* \supset A$ be sufficiently saturated, with $b \in M^*$. By Lemma 1.3.28, there is $p^*(x) \in S_\varphi(M^*)$, definable almost over A and containing $\varphi(x, b)$.

Let $\psi(y)$ be the definition of p^* which is a φ-formula almost over A. Let $q^* \in S_{\varphi^*}(M^*)$ be definable almost over A, consistent with $\mathrm{tp}(b/\mathrm{acl}^{eq}(A))$. Let $\tau(x)$ be the formula, over $\mathrm{acl}^{eq}(A)$, defining q^*.

By Remark 1.3.8, the formula $\tau(x)$ is a positive Boolean combination of $\mathrm{acl}^{eq}(A)$-conjugates of $\varphi(x, b)$. Notice that $\models \psi(b)$, so $\psi(y) \in q^*$, and by symmetry, we get $\tau(x) \in p^*$. So τ is consistent.

Let e, χ be such that $\tau(x) \leftrightarrow \chi(x, e)$, with $e \in \mathrm{acl}^{eq}(A)$. Let $e_1, \ldots, e_n$ be the realizations of $\mathrm{tp}(e/A)$. Then the formula $\chi(x, e_1) \vee \cdots \vee \chi(x, e_n)$ is over A, consistent, and is equivalent to a positive Boolean combination of $Aut(\bar{M}/A)$-conjugates of $\varphi(x, b)$, as τ is.

Conversely, let $\tau(x)$ be some consistent, over A, finite positive Boolean combination of $\mathrm{Aut}(\overline{M}/A)$ conjugates of $\varphi(x, b)$. Let $p(x) \in S_\varphi(A)$ contain $\tau(x)$. Let M^* be sufficiently saturated so that $\tau(x)$ is a positive Boolean combination of $\mathrm{Aut}(M^*/A)$ conjugates of $\varphi(x, b)$. Let $p'(x) \in S_\varphi(M^*)$ extend $p(x)$ and be definable over A.

So in particular, the formula $\tau(x)$ belongs to $p'(x)$. Since p' is a complete type, this implies that for some b' with $\mathrm{tp}(b'/A) = \mathrm{tp}(b/A)$, the formula $\varphi(x, b')$ belong to p'. Hence $\varphi(x, b')$ does not divide over A, and neither does $\varphi(x, b)$. □

We now are ready to prove Proposition 1.3.36:

Proof of Proposition 1.3.36. We will give a proof under a saturation assumption on M. The reader is invited to remove the assumption. $CB(-)$ denotes $CB_X(-)$ where $X = S_\varphi(\bar{M})$.

First, suppose that p' is definable almost over A. We can pick $\psi(x) \in p'(x)$ such that $R_\varphi(\psi) = R_\varphi(p')$.

Indeed, if not, for all formulas $\psi(x)$ contained in $p'(x)$, we would have $R_\varphi(\psi) > R_\varphi(p') = \mathrm{CB}(p')$. But because M is $(|A| + |T|)^+$-saturated, we have $R_\varphi(\psi) = \max\{\mathrm{CB}(q), q \in S_\varphi(M), \psi \in q\}$. Hence, for all formulas $\psi(x)$ contained in $p'(x)$, there is a type $q \in S_\varphi(M)$, containing $\psi(x)$, with $\mathrm{CB}(q) > \mathrm{CB}(p')$. Now, if we consider the open sets $\{[\psi(x)], \psi(x) \in p'(x)\}$, this implies, because ordinals are well ordered, that $\bigcap[\psi(x)]$ contains a type $q \in S_\varphi(\overline{M})$, with $\mathrm{CB}(q) > \mathrm{CB}(p')$. But this intersection is precisely $\{p'\}$, a contradiction.

By 1.3.37 and 1.3.28, some positive Boolean combination of A-conjugates of $\psi(x)$ is over A. Let $\psi_1(x) \vee \cdots \vee \psi_r(x)$ be this Boolean combination. Each of the $\psi_i(x)$ is consistent with $p(x)$, as $\psi(x)$ is. Hence, their disjunction also is consistent with $p(x)$. But since it is over A, it has to be equivalent to a formula in $p(x)$, say $\tau(x)$. In particular:

$$\begin{aligned} R_\varphi(p) &\leqslant R_\varphi(\tau) \\ &= R_\varphi(\psi_1(x) \vee \cdots \vee \psi_r(x)) \\ &= \max\{R_\varphi(\psi_i)\} \\ &= R_\varphi(\psi) \\ &= R_\varphi(p') \end{aligned}$$

so $R_\varphi(p) \leqslant R_\varphi(p')$. But we always have, if p' is an extension of p, that $R_\varphi(p) \geqslant R_\varphi(p')$ (to see this, use the restriction map $\pi : S_\varphi(M) \to S_\varphi(A)$ and an induction on CB rank). Hence $R_\varphi(p) = R_\varphi(p')$.

Conversely, suppose that $p'(x)$ is not definable almost over A. Let $\psi(y)$ be the definition of $p'(x)$, which is therefore not over $\mathrm{acl}^{eq}(A)$. In particular, by saturation of M, it has infinitely many images under $\mathrm{Aut}(M/A)$, giving rise to infinitely many distinct conjugate $p'_i \in S_\varphi(M)$ of p', all containing p.

This implies $\alpha = R_\varphi(p') < R_\varphi(p)$. Indeed, consider the compact set $[p(x)] \subset S_\varphi(M)$. This yields an infinite number of p'_i, all of CB rank α and contained in $[p(x)]$ (their CB rank is equal to their φ rank by saturation of M). Hence, the set $\{p'_i, i \in I\}$ must have an accumulation point q in $[p(x)]$. And because they all have rank α, we get $CB(q) > \alpha$. But we also have $CB(q) < CB(p)$, as q extends p, hence $R_\varphi(p) = CB(p) > \alpha = R_\varphi(p')$. □

This concludes our exposition of local (formula-by-formula) stability. In the next section we will apply this to obtain information about stable theories.

1.3.2 *Stable theories*

A theory T is said to be stable if every formula $\varphi(x, y)$ is stable for T. Unless otherwise stated, in this section we will assume that the theory T is stable, and that $\overline{M} \models T$ is some large, sufficiently saturated and homogeneous model.

Let $\Delta = \{\varphi_1(x, y_1), \dots, \varphi_n(x, y_n)\}$ be a finite set of stable formulas. By a Δ-formula, we mean a Boolean combination of instances of $\varphi_i(x, y_i)$, $1 \leqslant i \leqslant n$. For some set A, a complete Δ-type over A, $p(x) \in S_\Delta(A)$, is a maximal, consistent set of Δ-formulas *over* A (i.e. Δ-formulas equivalent modulo T to a formula over A). As usual, if $A = M \models T$, then $p(x) \in S_\Delta(M)$ is determined by the instances of $\varphi_i(x, y_i)$ and $\neg\varphi_i(x, y_i)$ (with parameters in M) that appear in $p(x)$. Everything from Section 1.3.1 holds for Δ-formulas and Δ-types. In particular, we have

Lemma 1.3.38. *Let $\Delta(x) = \{\varphi_1(x, y_1), \dots, \varphi_n(x, y_n)\}$ be a finite set of stable formulas.*

(i) Any $p(x) \in S_\Delta(M)$ is definable (i.e. every $\varphi_i(x, y_i)$ has a defining formula).
(ii) Suppose $q(x) \in S_x(A)$ is a type, $A \subseteq M \models T$. Then there is $p(x) \in S_\Delta(M)$ such that $p(x)$ is definable almost over A and $p(x) \cup q(x)$ is consistent.

From this, we can conclude the following:

Proposition 1.3.39. *Let T be a stable theory, and let $A = \operatorname{acl}^{eq}(A) \subseteq \overline{M} \models T$ and $p(x) \in S_x(A)$. Then, for any $M \supseteq A$ (in particular, for M sufficiently saturated), there is $p'(x) \in S_x(M)$ such that $p(x) \subseteq p'(x)$ and $p'(x)$ is definable over A. Moreover, $p'(x)$ is the unique such type.*

Proof. Fix $M \supseteq A$. By Lemma 1.3.15 and Lemma 1.3.20, for every formula $\varphi(x, y)$, there is a unique φ-type $p'_\varphi(x) \in S_\varphi(M)$ that is consistent with $p(x)$ and definable over A. Consider the set of formulas

$$p'(x) = \bigcup_{\varphi(x,y)} p'_\varphi(x).$$

We claim that $p'(x)$ is consistent. By compactness, it is enough to show that for any finite set of formulas $\Delta = \{\varphi_1(x, y_1), \ldots, \varphi_n(x, y_n)\}$, the set

$$p'_\Delta(x) = \bigcup_{i=1}^{n} p'_{\varphi_i}(x)$$

is consistent. By Lemma 1.3.38*(ii)* there is a Δ-type $q(x) \in S_\Delta(M)$ such that $p(x) \cup q(x)$ is consistent, and such that $q(x)$ is definable over A. By the uniqueness of p'_{φ_i}, we get $q \upharpoonright \varphi_i = p'_{\varphi_i}$ for $1 \leqslant i \leqslant n$, and hence $p'_\Delta(x) \subset q(x)$ and so $p'_\Delta(x)$ is consistent.

Maximality is because each p'_φ is a complete φ-type, and uniqueness comes from the uniqueness of the p'_φ. □

Proposition 1.3.40. *Let $p(x) \in S_x(A)$, $q(x) \in S_x(B)$ and $p(x) \subseteq q(x)$. The following are equivalent:*

(i) there is $M \supseteq B$ and $q'(x) \in S_x(M)$ such that $q'(x) \supseteq q(x)$ and $q'(x)$ is definable almost over A,

(ii) $q(x)$ does not fork over A.

Proof. Assume *(i)*. Fix any formula $\varphi(x, y) \in L$. So the restriction $q'(x) \upharpoonright \varphi$ to a complete φ-type over M' is definable almost over A. By Corollary 1.3.33, $q' \upharpoonright \varphi$ does not fork over A, so $q \upharpoonright \varphi$ does not fork over A. By varying φ we see that every formula in q dies not fork over A, giving *(ii)*

For the other direction, assume that $q(x)$ does not fork over A. Let $M \supseteq B$ be arbitrary. By Remark 1.3.31 q extends to a complete type q' over M which does not divide over A. Restricting to arbitrary formulas $\varphi(x, y)$ and using again Corollary 1.3.33, we conclude that q' is definable almost over A. □

Note that by 1.3.32 we have:

Remark 1.3.41. Suppose T is stable and $A \subseteq B$. Then $q(x) \in S(B)$ forks over A if and only if $q(x)$ divides over A.

Here we introduce Makkai's anchor notation for independence. We write $\bar{a} \mathop{\smile\!\!\!\!\!\mid}_B C$ to mean that $\text{tp}(\bar{a}/BC)$ does not fork over B. We extend this notation to sets by writing $A \mathop{\smile\!\!\!\!\!\mid}_B C$ to mean $\bar{a} \mathop{\smile\!\!\!\!\!\mid}_B C$ for every finite tuple $\bar{a} \in A$.

Proposition 1.3.42 (Properties of $\mathop{\smile\!\!\!\!\!\mid}$/forking). *Let T be a stable theory.*

1. *(Existence) Let $p(x) \in S_x(A)$ and $A \subseteq B$. Then there exists $q(x) \in S_x(B)$, with $q(x) \supseteq p(x)$ and $q(x)$ does not fork over A.*

2. *(Transitivity) Let* $A \subseteq B \subseteq C$ *and* $p(x) \in S_x(A)$, $q(x) \in S_x(B)$, $r(x) \in S_x(C)$, *with* $p(x) \subseteq q(x) \subseteq r(x)$. *Then* $r(x)$ *does not fork over* A *iff* $r(x)$ *does not fork over* B *and* $q(x)$ *does not fork over* A.
3. *(Symmetry) Given* $\bar{a}$, *and* $A \subseteq B$, $\mathrm{tp}(\bar{a}/B)$ *does not fork over* A *iff* $\mathrm{tp}(\bar{b}/A\bar{a})$ *does not fork over* A *for all finite* $\bar{b}$ *from* B.
4. *(Finite character) Let* $p(x) \in S_x(B)$ *and* $A \subseteq B$. *Then* p *does not fork over* A *if and only if for each finite tuple* b *from* B, $p\restriction(A, b)$ *does not fork over* A.
5. *(Local Character) For all* $q(x) \in S_x(B)$, *there is* $A \subseteq B$, $|A| \leqslant |T|$ *such that* $q(x)$ *does not fork over* A.
6. *(Uniqueness) If* $A = \mathrm{acl}^{eq}(A)$ *and* $p(x) \in S_x(A)$, *then for all* $B \supseteq A$ *there is unique* $q(x) \supseteq p(x)$ *that does not fork over* A.
7. *(Algebraicity) If* $q(x) \in S_x(B)$ *is algebraic (i.e. has only finitely many realizations) and* $q(x)$ *does not fork over* A, *then* $q\restriction_A$ *is algebraic.*
8. *(Conjugacy and the finite equivalence relation theorem) If* $p(x) \in S_x(A)$ *and* $B \supseteq A$, *then* p *has at most* $2^{|T|}$ *many non-forking extensions* $q(x) \in S_x(B)$. *Moreover, if* $B = M \models T$ *is* $|T| + |A|$*-saturated and strongly homogeneous, then* $\mathrm{Aut}(M/A)$ *acts transitively on the set of non-forking extensions* $q(x) \in S_x(M)$, *and if* q_1 *and* q_2 *are distinct such non-forking extensions of* p, *there is an* A*-definable finite equivalence relation* $E(x_1, x_2)$ *such that* $q_1(x_1) \cup q_2(x_2) \vdash \neg E(x_1, x_2)$.

We can restate these properties 1–7 using the anchor notation as follows:

1. *(Existence) For all* $A \subset B$ *and tuple* e, *there a tuple* e' *satisfying both* $\mathrm{tp}(e/A) = \mathrm{tp}(e'/A)$ *and* $e \mathop{\smile\!\!\!\!\!|}_A B$.
2. *(Transitivity) For all* A, B, C, D, *we have* $A \mathop{\smile\!\!\!\!\!|}_B CD$ *if and only if* $A \mathop{\smile\!\!\!\!\!|}_B C$ *and* $A \mathop{\smile\!\!\!\!\!|}_{BC} D$.
3. *(Symmetry)* $A \mathop{\smile\!\!\!\!\!|}_C B$ *if and only if* $B \mathop{\smile\!\!\!\!\!|}_C A$.
4. *(Finite character)* $A \mathop{\smile\!\!\!\!\!|}_C B$ *iff* $A_0 \mathop{\smile\!\!\!\!\!|}_C B_0$ *for all finite* $A_0 \subseteq A$ *and finite* $B_0 \subseteq B$.
5. *(Local Character) For all finite* $\bar{a}$, *if* B *is any set, there is* $C \subseteq B$ *with* $|C| \leqslant |T|$ *such that* $\bar{a} \mathop{\smile\!\!\!\!\!|}_C B$.
6. *(Uniqueness) If* $A = \mathrm{acl}^{eq}(A)$, $\mathrm{tp}(a_1/A) = \mathrm{tp}(a_2/A)$, $a_1 \mathop{\smile\!\!\!\!\!|}_A B$, *and* $a_2 \mathop{\smile\!\!\!\!\!|}_A B$, *then* $\mathrm{tp}(a_1/B) = \mathrm{tp}(a_2/B)$.
7. *(Algebraicity) If* $A \mathop{\smile\!\!\!\!\!|}_B C$ *and* $A \subseteq acl^{eq}(B, C)$ *then* $A \subseteq acl^{eq}(B)$.

Proof. The proof follows from the formula-by-formula case as well as the definitions. □

Remark 1.3.43.

1. The properties of $\mathop{\smile\!\!\!\!\!\!\mid}$, other than uniqueness, characterize a broader class of theories called simple theories. In fact, the local character property can be taken as a definition of simple theories. Examples of simple theories include the theory of the random graph, the theory of pseudofinite fields, and the model companion to the theory of algebraically closed fields with an automorphism (ACFA).
2. Uniqueness is characteristic of stable theories in the following sense: Let $A \subseteq B$ and let $\mathcal{F}_A$ be the set of formulas $\varphi(x)$ over B that fork over A. Then $\mathcal{F}_A$ is a proper ideal in the Boolean algebra of formulas over B (if $\varphi(x) \vee \psi(x) \in \mathcal{F}_A$, then $\varphi(x) \in \mathcal{F}_A$ or $\psi(x) \in \mathcal{F}_A$). Furthermore, if $A = \mathrm{acl}^{eq}(A)$ and $p(x) \in S_x(A)$, then there is a unique $q(x) \in S_x(B)$ extending $p(x)$ and avoiding $\mathcal{F}_A$, and these properties characterize T being stable.
3. In many examples of stable theories, $\mathop{\smile\!\!\!\!\!\!\mid}$ has a natural interpretation. For example, in ACF, $\bar{a} \mathop{\smile\!\!\!\!\!\!\mid}_k \bar{b}$ iff $\mathrm{tr.deg.}(k(\bar{a})/k) = \mathrm{tr.deg.}(k(\bar{a},\bar{b})/k(\bar{b}))$. There is also a more geometrical interpretation: let k be any field (not necessarily algebraically closed) and let V be an irreducible k-variety. There is a complete ACF type $p_V(\bar{x})$, which says that $\bar{x} \in U$ for every Zariski open $U \subseteq V$ defined over k. Then for any field $F \supseteq k$, then $p_V(\bar{x})$ has a unique non-forking extension to F if and only if V remains irreducible as a variety over F. Note then that $p_V(\bar{x})$ has a unique non-forking extension to k^{alg} if and only if V is absolutely irreducible.
4. Let T be stable, $\overline{M} \models T$, and $A \subseteq \overline{M}$ a small set. We present some examples of "forking calculus".

   (i) Suppose that $(a_\alpha : \alpha < \kappa)$ are tuples in $\overline{M}$ such that $a_\alpha \mathop{\smile\!\!\!\!\!\!\mid}_A \{a_\beta : \beta < \alpha\}$ for all $\alpha < \kappa$. Then $a_\alpha \mathop{\smile\!\!\!\!\!\!\mid}_A \{a_\beta : \alpha \neq \beta < \kappa\}$ and we say that $\{a_\alpha : \alpha < \kappa\}$ is an A-independent set.

   Proof. By finite character, it is enough to prove, for any $\alpha < \kappa$, any n and $\beta_1, \cdots, \beta_n < \kappa$, that $a_\alpha \mathop{\smile\!\!\!\!\!\!\mid}_A a_{\beta_1} \cdots a_{\beta_n}$. We will do so by induction on n. If $n = 1$, it is an immediate consequence of the assumption.

   Now assume $n > 1$. If $\alpha > \beta_n$ for all n, this is the assumption. Otherwise, we can assume, by reordering the

β_i, that if $i < j$ then $\beta_i < \beta_j$. By assumption, we get $a_{\beta_n} \mathop{\smile\!\!\!\!\!\!|}_A a_\alpha a_{\beta_1}, \cdots a_{\beta_{n-1}}$. Applying transitivity, this yields $a_{\beta_n} \mathop{\smile\!\!\!\!\!\!|}_{A a_{\beta_1} \cdots a_{\beta_{n-1}}} a_\alpha$, and by the induction hypothesis (and transitivity) we get $a_{\beta_n} \cdots a_{\beta_1} \mathop{\smile\!\!\!\!\!\!|}_A a_\alpha$, what we wanted to prove. □

(ii) (Weight) Suppose that $\{a_\alpha : \alpha < |T|^+\}$ is an A-independent set of finite tuples. Then for any b, there is α such that $b \mathop{\smile\!\!\!\!\!\!|}_A a_\alpha$.

Proof. For a contradiction, suppose that there is b such that $b \not\mathop{\smile\!\!\!\!\!\!|}_A a_\alpha$ for all $\alpha < |T|^+$. We claim that for all $\alpha < |T|^+$, $b \not\mathop{\smile\!\!\!\!\!\!|}_{A \cup \{a_\beta : \beta < \alpha\}} a_\alpha$. To see this, fix some $\alpha < |T|^+$. By assumption, $a_\alpha \mathop{\smile\!\!\!\!\!\!|}_A \{a_\beta : \beta < \alpha\}$. As $a_\alpha \not\mathop{\smile\!\!\!\!\!\!|}_A b$, it follows from transitivity that $a_\alpha \not\mathop{\smile\!\!\!\!\!\!|}_{A \cup \{a_\beta : \beta < \alpha\}} b$. By symmetry, we have that $b \not\mathop{\smile\!\!\!\!\!\!|}_{A \cup \{a_\beta : \beta < \alpha\}} a_\alpha$, as required.
Now, by local character, there is $B_0 \subseteq \{a_\alpha : \alpha < |T|^+\}$ such that $b \mathop{\smile\!\!\!\!\!\!|}_{AB_0} \{a_\alpha : \alpha < |T|^+\}$ and $|B_0| \leqslant |T|$. As $|T|^+$ is regular, there is $\gamma < |T|^+$ such that $B_0 \subseteq \{a_\alpha : \alpha < \gamma\}$. But then $b \mathop{\smile\!\!\!\!\!\!|}_{AB_0} a_\gamma$, contradicting the fact that for all $\alpha < |T|^+$, $b \not\mathop{\smile\!\!\!\!\!\!|}_{A \cup \{a_\beta : \beta < \alpha\}} a_\alpha$. □

Denote $p = \text{tp}(b/A)$. By what we just proved, the size of a cardinal α such that there is an A-independent set of realizations of p of size α, is bounded by $|T|^+$. Hence, we can take the supremum of such cardinals, it is called the preweight of the type p.
From this, we define the weight of p as the supremum of the preweights of non forking extensions of p, it is also a well-defined cardinal.

1.3.3 *A survey of classification theory*

In addition to Shelah's book [She90], the reader can consult John Baldwin's book [Bal88] for proper and detailed references for facts and results stated in this section.

Classification theory, in the sense of Shelah, is concerned with finding meaningful dividing lines among complete first order theories. In earlier times, the decidability/undecidability dichotomy was considered the fundamental dividing line, but it is no longer the case, as dividing lines of a different nature have become as or more important. However, decidability is still very interesting and can sometimes be deduced from structural properties such as quantifier elimination.

In contemporary model theory, there are many different crosscutting notions of complexity, some of which give dividing lines, some of which just give interesting properties.

One such measure of complexity is called the spectrum of a theory T, denoted by $I(\kappa, T)$. Assume that T is a complete, countable theory with no finite models. Then $I(\kappa, T)$ is defined to be the cardinality of the set of isomorphism classes of models of cardinality κ. In 1965, Morley proved his well-known categoricity theorem, that is, if $I(\kappa, T) = 1$ for some uncountable κ, then $I(\kappa, T) = 1$ for all uncountable κ.

In the case of $\kappa = \aleph_0$, much less is known. It is a famous conjecture by Vaught that either $I(\aleph_0, T) \leqslant \aleph_0$ or $I(\aleph_0, T) = 2^{\aleph_0}$ (without assuming the continuum hypothesis, of course).

In the early eighties, Shelah proved his "main gap" conjecture. To state it, we need to recall some set theoretic notation:

Definition 1.3.44. We define, for any ordinal α, the cardinal $\beth_\alpha$ by induction:

- $\beth_0 = \aleph_0$
- if $\alpha = \beta + 1$, then $\beth_\alpha = 2^{\beth_\beta}$
- if α is limit, then $\beth_\alpha = \sup\{\beth_\beta, \beta < \alpha\}$.

The main gap theorem then states that for all ordinals $\alpha > 0$, either $I(\aleph_\alpha, T) = 2^\kappa$ for all $\alpha \geqslant 1$, or $I(\aleph_\alpha, T) < \beth_{\omega_1}(|\omega + \alpha|)$. The idea is that either a theory T has a maximal number of models, or there is some kind of classification of the models of T. One reason for studying stable theories is the following theorem due to Shelah:

Theorem 1.3.45. *For all unstable T and all uncountable κ, $I(\kappa, T) = 2^\kappa$.*

So, from the point of view of $I(\kappa, T)$ as a measure of complexity, we may assume that T is stable (note that when $\kappa = \aleph_0$, there is no analogous result; the theory of dense linear orders is unstable, but is $\aleph_0$-categorical).

Definition 1.3.46. Let T be a one-sorted theory. We say T is strongly minimal if, for any formula $\varphi(x)$ in one variable and any model M of T, $\varphi(M)$ is either finite or cofinite.

Note that every strongly minimal theory is stable: Suppose T is strongly minimal, and that M is a countable model. Then there are only countably many 1-types: the algebraic types, i.e. those which contain (and are isolated by) a formula with only finitely many realizations, and the unique (unrealized) type $p(x)$ expressing that x is in every cofinite set.

Example 1.3.47. The following examples can be shown to be strongly minimal via quantifier-elimination (note that quantifier-elimination does not imply strong minimality in general, it just makes it easier to study the formulas in one variable):

(i) the theory of an infinite set in the language of equality,
(ii) the theory of an infinite vector space over a countable field F in the language of modules (i.e. the language of groups together with a function symbol λ_r for scalar multiplication by r for all $r \in F$),
(iii) algebraically closed fields,
(iv) $\mathrm{Th}(\mathbb{Z}_{p^\infty})$, the theory of the Prüfer p-group.

Fact 1.3.48. *Suppose T is strongly minimal, $M \models T$, and $A \subseteq M$. Then the algebraic closure over A in M is a pregeometry, i.e. for $a, b \in M$ if $b \in \mathrm{acl}(A, a) \setminus \mathrm{acl}(A)$, then $a \in \mathrm{acl}(A, b)$.*

Definition 1.3.49. A tuple $a_1, \cdots, a_n$ is said to be A-algebraically independent if for any i, we have $a_i \notin \mathrm{acl}(A \cup (\{a_1, \cdots a_n\} \setminus \{a_i\}))$.

Note that by properties of algebraic closure, if $b_i \in \mathrm{acl}(A, \overline{a})$ for all $i = 1, \ldots, n$ and $c \in \mathrm{acl}(b_1, \ldots, b_n, \overline{a}, A)$, then $c \in \mathrm{acl}(A, \overline{a})$. This, and the previous fact, allow us to define, for $\overline{b}$ a finite tuple from $\overline{M}$, the dimension of $\overline{b}$ over A, denoted $\dim(\overline{b}/A)$, as the maximal size of an A-algebraically independent subtuple of $\overline{b}$.

With a bit of forking calculus, the reader can prove that if $A \subset B$, then $\overline{b} \mathop{\smile\!\!\!\!\!\mid}_A B \Leftrightarrow \dim(\overline{b}/B) = \dim(\overline{b}/A)$.

Fact 1.3.50. *Let T be strongly minimal. Then*

$$I(\kappa, T) = \begin{cases} 1, & \text{if } \kappa > \omega, \\ 0 \text{ or } \aleph_0, & \text{if } \kappa = \aleph_0. \end{cases}$$

Proof. (For when $\kappa > \omega$) Let $M \models T$ and $I \subseteq M$ be a maximal algebraically independent set over $\varnothing$. Then $|I|$ is well-defined and $M = \mathrm{acl}(I)$. Moreover, the type $\mathrm{tp}(I/\varnothing)$ is determined by $|I|$. If $|M| = \kappa > \omega$, then $|I| = \kappa$. If $|M| = \aleph_0$ however, there is a bit more work to do. □

Definition 1.3.51. T is *ω-stable* if for every countable $M \models T$ and finite tuple of variables x, $|S_x(M)| = \aleph_0$.

Note that strongly minimal implies ω-stable, which in turn implies stable.

Fact 1.3.52. *T is ω-stable if and only for any finite tuple of variables x and $X = S_x(\bar{M})$, every $p \in X$ has ordinal valued Cantor–Bendixson rank. In that case, the Cantor–Bendixson rank is better know as* $\mathrm{RM}(p)$*, the Morley rank.*

We can, as was done in Definition 1.3.35, define the Morley rank of a formula.

Note.
(i) If T is strongly minimal then $RM(x = x) = 1$.
(ii) (Assuming ω-stability.) Given a formula $\varphi(x)$, there are only finitely many $p \in S_x(\overline{M})$ which containing $\varphi(x)$ with maximal Morley rank. This finite number is called the Morley degree $\mathrm{dM}(\varphi(x))$ of φ.

A lot of natural and interesting theories are ω-stable, let us give a few examples.

Example 1.3.53. The theory DCF_0 of differentially closed fields of characteristic zero, is ω-stable (see 1.3.68 for a definition). Moreover, the formula $x = x$ has Morley rank ω.

ω-stable theories can have many uncountable models (2^κ models of cardinality κ for all uncountable κ) as is the case for DCF_0 or can have a structure theory for their (uncountable) models as in the following example.

Example 1.3.54. The theory $T = \mathrm{Th}\left(\mathbb{Z}_{p^\infty}^{(\omega)}, +\right)$ is ω-stable of Morley rank ω. Moreover, the models of T are precisely given by $\mathbb{Z}_{p^\omega}^{(\kappa)} \oplus \mathbb{Q}^{(\lambda)}$, where $\kappa > \omega$ and $\lambda > 0$.

A useful property of ω-stable theories is that there are prime models over all sets, i.e. for any $M \models T$ and $A \subseteq M$, the theory $\mathrm{Th}(M, a)_{a \in A}$ has a unique prime model. Also Vaught's Conjecture is known for ω-stable theories T.

Another class of stable theories, slightly more general than ω-stable ones, are the superstable theories.

Definition 1.3.55. T is *superstable* if T is stable and for every $p(x) \in S_x(B)$ there is a finite set $A \subseteq B$ such that $p(x)$ does not fork over A.

Fact 1.3.56.

1. *T is ω-stable if and only if*
   - *T is superstable*
   - *$S(T)$ $(= \bigcup_x S_x(T))$ is countable*
   - *every $p(x) \in S(B)$ has finite multiplicity (i.e. has only finitely many non-forking extensions to a given $M \supseteq B$).*
2. *If T is not superstable, then $I(\kappa, T) = 2^\kappa$ for all $\kappa > \omega$.*
3. *Vaught's Conjecture is still open for superstable theories.*

Example 1.3.57. $\mathrm{Th}(\mathbb{Z}, +, 0)$ is superstable.

Exercise. If $M = (\mathbb{Z}, +, 0)$, prove that $|S_1(M)| = 2^{\aleph_0}$.

Solution. Since the language is countable, $S_1(M) \leqslant 2^{\aleph_0}$. For each $i \in \omega$, define

$$\begin{aligned}\varphi_i(x) &:= \text{“}x \text{ is a multiple of the } i^{\text{th}} \text{ prime”} \\ &= (\exists y)\,[\underbrace{y + \ldots + y}_{i^{\text{th}} \text{ prime times}} = x].\end{aligned}$$

Then for each $\sigma \in 2^{<\omega}$, define $p_\sigma(x) := \{\varphi_i(x) : \sigma(i) = 1\} \cup \{\neg\varphi_i(x) : \sigma(i) = 0\}$. This definition can be extended in the natural way to define $p_f(x)$ for each $f \in 2^\omega$. Now each $p_\sigma(x)$ is satisfiable, thus by the compactness theorem, each $p_f(x)$ is also satisfiable. Therefore, by taking a completion of each $p_f(x)$, there are at least $2^{\aleph_0}$ many 1-types given by $\{p_f(x) : f \in 2^\omega\}$.

Even though this theory is not ω-stable, there is a description of the models. Let $\hat{\mathbb{Z}}$ be the profinite completion of $\mathbb{Z}$, i.e. $\hat{\mathbb{Z}}$ is the inverse (or projective) limit of the $\mathbb{Z}/n\mathbb{Z}$.

Then the $2^{\aleph_0}$-saturated models of T are precisely of the form

$$\hat{\mathbb{Z}} \oplus \mathbb{Q}^{[\delta]}, \quad \delta \geqslant 2^{\aleph_0}.$$

In fact the models of T are precisely of the form

$$G \oplus \mathbb{Q}^{[\delta]}, \quad \delta \geqslant 2^{\aleph_0}$$

for some $G < \hat{\mathbb{Z}}$.

Example 1.3.58. The following theories are stable but not superstable.

- $\mathrm{Th}(\mathbb{Z}, +)^{(\omega)}$

- $\text{Th}((\mathbb{F}_p)^{sep}, +, \times)$
- $\text{Th}(F_2, \cdot)$, where F_2 is the free group on 2 generators.

Remark 1.3.59. The last statement was proved by Zlil Sela, and opened a whole new area of exploration for model theory.

To prove nonsuperstability in the first two cases, one can use the following general statement: if T is a theory, G a group definable in a model M of T, and there is a sequence $G = G_0 \geqslant G_1 \geqslant G_2 \geqslant \cdots$ of definable subgroups where G_{i+1} has infinite index in G_i for all i, then T is not superstable.

Many theories that are interesting from a model theoretic perspective are actually unstable. For example, the theory of the reals $\text{Th}(\mathbb{R}, +, \times)$ and the theory of the p-adics $\text{Th}(\mathbb{Q}_p,, \times)$ are unstable.

These would be the subject of another course, but here, we have restricted our attention to stable theories. Therefore, let us fix (again) a stable theory T.

Definition 1.3.60. $p(x) \in S(A)$ is stationary if for all $B \supseteq A$ there is a unique nonforking extension $q(x) \in S(B)$ of p. For example, if $A = \text{acl}^{\text{eq}}(A)$, then any $p(x) \in S(A)$ is stationary.

Lemma 1.3.61. *Let $p(x) \in S(A)$ be stationary. Then there exists a unique smallest $A_0 \subseteq \text{dcl}^{eq}(A)$ such that $p(x)$ does not fork over A_0, and $p{\restriction}A_0$ is stationary. This set A_0 is denoted* $\text{Cb}(p)$, *and called the canonical base of p.*

Proof. (Sketch) Let $M \supseteq A$, and let $q(x)$ be the nonforking extension of $p(x)$ to M. Then $q(x)$ is definable by stability. For each $\varphi(x, y) \in L$, let $\psi_\varphi(y)$ be a formula over M which is the φ-definition of p. Then $A_0 = \text{dcl}\{\text{codes of } \psi_\varphi(y) : \varphi \in L\}$. □

1.3.4 *Geometric stability theory*

For this section, see [Pil96] and [Be98] for background and references to original papers.

As mentioned earlier, there are many notions of complexity in model theory, for example coming from decidability/undecidability, or coming from the stability hierarchy: ω-stable, superstable, stable unsuperstable, unstable.

In geometric stability theory, we deal with another kind of complexity: the behaviour of definable families of definable subsets of a given ambient space, in particular how these definable subsets may intersect each other. At one extreme is the family of translates (cosets) of H in G where H is a definable subgroup of a definable group G. These cosets are equal or disjoint. At the other extreme is the family of lines

$$\{y = ax + b : a, b \in \mathbb{C}\} \subseteq \mathbb{C}^2$$

which is a definable family of definable sets in the complex field. We can recover the field from the incidence system consisting of $\mathbb{C}^2$ and the family of lines.

This kind of "geometric" complexity is also bound up with whether or not infinite groups are definable in a structure, and if so what kind of groups (abelian, simple, ...). Also whether or not infinite fields are definable.

Let T be a stable theory, and X a definable set in a saturated model $\overline{M}$ of T. We call X strongly minimal (with respect to the ambient T or $\overline{M}$) if X is infinite and every definable (with parameters in $\bar{M}$) subset of X is finite or cofinite. Equivalently, $\mathrm{RM}(X) = \mathrm{dM}(X) = 1$.

Given any definable (without parameters) set X in $\overline{M}$, by X^{eq}, we mean the collection of all sorts S_E of T^{eq} (or $\overline{M}^{\mathrm{eq}}$) where E is a $\emptyset$-definable equivalence relation on $X \times \ldots \times X$ (n times for some n).

Zilber suggested that we could classify or describe strongly minimal sets, and that moreover, that any "rich enough" strongly minimal set arises from an algebraic object:

Conjecture. *(Zilber) Let X be a strongly minimal set (in $\overline{M} \models T$) definable without parameters. Then exactly one of the following holds:*

1. *The algebraic closure in X is "trivial" i.e. for $a_1, \ldots, a_n \in X$,*

$$\mathrm{acl}(a_1, \ldots, a_n) \cap X = \bigcup_{i=1}^{n} (\mathrm{acl}(a_i) \cap X).$$

2. *There is a strongly minimal group G definable in X^{eq}, and the group G has the property that any definable (with parameters) subset of $G \times \cdots \times G$ is a Boolean combination of cosets of definable subgroups of $G \times \cdots \times G$.*
3. *There is a strongly minimal field $(K, +, \times)$ definable in X^{eq}.*

These three alternatives measure the complexity of the strongly minimal set, from simplest to most complex. The Zilber conjecture was proven to be false. Nonetheless, it has been extremely influential, and we will discuss here some of the most important ideas it led to. First, let us define:

Definition 1.3.62. Let X be any definable set, say defined without parameters. X is said to be one based if for any tuple $\bar{a}$ from X, and any $B = \mathrm{acl}^{eq}(B)$, we have $\mathrm{Cb}(\mathrm{tp}(\bar{a}/B)) \in \mathrm{acl}(\bar{a})$. A one-sorted theory T is one-based if the underlying set (defined by $x = x$) is one-based.

A general result in stability theory, which will not prove here is:

Proposition 1.3.63. *Let $p \in S(B)$ be a stationary type, and $(a_i)_{i \in \mathbb{N}}$ be a Morley sequence in p, namely an independent over B sequence of realizations of p. Then* $\mathrm{Cb}(p) \in \mathrm{acl}(\{a_i, i \in \mathbb{N}\})$.

The example of the family of lines in $\mathbb{C} \times \mathbb{C}$ yields non 1-basedness of the theory ACF_0.

On the other hand quantifier elimination shows that the theory of infinite dimensional vector spaces over a given field F (in the F-module language) is one-based.

This definition is linked with Zilber's conjecture by the following:

Theorem 1.3.64. *Let X be strongly minimal. Then X is one based if and only if case 1 or 2 hold.*

This is nontrivial, and one of the early result in geometric stability theory.

As we mentioned, the trichotomy is false. In fact, it fails in a very strong sense, as proved by Hrushovski [Hru93].

Theorem 1.3.65. *There is a strongly minimal set (in fact theory) which is not one-based, but does not interpret any infinite group.*

Note that this indeed disproves the conjecture by Theorem 1.3.64, as such a set would have to satisfy alternative 3 of Zilber's trichotomy. But if it did, it would interpret an algebraically field, and this structure does not even interpret an infinite group!

To construct this strongly minimal set, a variant of Fraissé amalgamation was used, now known as a Hrushovski construction.

Nevertheless the truth of the Zilber conjecture in natural theories has had striking consequences.

One of the most important example is DCF_0, the theory of differentially closed fields of characteristic zero, which we will now define.

Let $\mathcal{L}$ be the language of rings, with one additional function symbol ∂. A differential ring R is an $\mathcal{L}$-structure satisfying the ring axioms, and such that ∂ is a derivation, i.e. for all $a, b \in R$, we have $\partial(a+b) = \partial(a) + \partial(b)$ and $\partial(ab) = \partial(a)b + a\partial(b)$.

The theory of differential fields of characteristic zero has a model companion, which is the theory of differentially closed fields, denoted DCF_0.

Definition 1.3.66. Let (K, ∂_0) be a differential field. The ring $K\{x\}$ of differential polynomials over K is defined as the polynomial ring $K[\{\partial^i(x), i \in \mathbb{N}\}]$, with the differential ring structure given by $\partial = \partial_0$ on K, and $\partial(\partial^{(i)}x)) = \partial^{(i+1)}(x)$.

Definition 1.3.67. Let f be a differential polynomial. Then the order of f is the biggest i such that $\partial^{(i)}(x)$ appears in f.

Definition 1.3.68. The theory DCF_0 is axiomatized by the following:

- axioms forACF_0
- for any differential polynomials f, g, with the order of g strictly larger than the order of f, there is a such that $g(a) = 0$ and $f(a) \neq 0$.

Equipped with these axioms, one can prove the following basic properties of DCF_0:

- quantifier elimination
- elimination of imaginaries
- ω-stable of Morley rank ω.

In particular, any definable set will be given as the solution of differential equations and inequations.

Moreover, Zilber's trichotomy is true in DCF_0. This was first proven by Hrushovski and Sokolovic, using the machinery of Zariski geometries (Hrushovski–Zilber), which we will not describe here, but see [Be98] for details and references. Later, Pillay and Ziegler, in [PZ03], obtained a second proof, as a corollary of proving a strong structural property of differentially closed fields, called the canonical base property.

Let us now fix a monster model $\mathbb{M}$ of DCF_0. Let $\mathcal{C} = \{x \in \mathbb{M}, \partial(x) = 0\}$ be the field of constants of $\mathbb{M}$. This is an algebraically closed field, and moreover, any definable subset (possibly with parameters) of $\mathcal{C}$ is already definable in $(\mathcal{C}, +, \times)$, with parameters from $\mathcal{C}$. We say that $\mathcal{C}$ is a purely stably embedded algebraically closed field.

Let us now give a more explicit version of the trichotomy in DCF_0. In case one, nothing more can be said about the strongly minimal set X.

However, identifying exactly what differential equations give rises to trivial minimal sets is an ongoing and active project.

In the third case, the algebraically closed field definable in the set X can be proven to be definably isomorphic to $\mathcal{C}$.

The second and third cases are a bit complicated and we will say now something about the second case. To discuss it, we need to define some basic notions from algebraic geometry. For a more detailed introduction to this subject, we refer the reader to [SR94]. Let K be an algebraically closed field (of characteristic 0 if one wishes), which we consider as a structure in the language of rings.

Definition 1.3.69. An affine algebraic variety is a subset of K^n, for some n, defined as the zero locus of some finite system of polynomial equations.

Definition 1.3.70. Projective n-space, denoted by $\mathbb{P}^n(K)$, is defined as $K^n \backslash \{0\}$, quotiented by the equivalence relation $\bar{a} \sim \bar{b}$ if and only if there is $\lambda \in K\backslash\{0\}$ such that $\bar{a} = \lambda \cdot \bar{b}$.

A projective algebraic variety is a subset of $\mathbb{P}^n(K)$ defined as the zero locus of a finite system of homogeneous polynomials. Note that this zero set is well defined because if P is an homogeneous polynomial, then there exist $k \in \mathbb{N}$ such that for all $\bar{a}, \lambda$, we have $P(\lambda \cdot \bar{a}) = \lambda^k \cdot P(\bar{a})$.

Morphisms between affine algebraic varieties are given by polynomial maps. Morphisms of projective algebraic varieties are basically given by systems of homogeneous polynomials.

Definition 1.3.71. An abelian variety is an irreducible projective variety G equipped with a group operation $m : G \times G \to G$ which is a morphism of projective varieties.

Definition 1.3.72. An abelian variety is simple if it has no proper non-trivial abelian subvariety.

It can be showed that all abelian varieties are commutative groups. The following will be of importance:

Definition 1.3.73. Let A be an abelian variety defined over K, and $k \subset K$ another algebraically closed field. We say that A descends to k is there is an abelian variety A_0, defined over k, which is definably isomorphic (as a group) to A.

Let A be an abelian variety defined in our differentially closed field $\mathbb{M}$. Its Morley rank can be computed as $\mathrm{RM}(A) = \omega \cdot \dim(A)$, where $\dim(A)$

is the dimension of A as an algebraic variety. Note that A can indeed be viewed as a definable set, by elimination of imaginaries.

Fact. *A has a unique smallest Zariski-dense definable in $(\mathbb{M}, +, \times, \partial)$ subgroup $A^{\#}$. If A is simple, then $A^{\#}$ is strongly minimal.*

We can now states the second alternative of the trichotomy. If X interprets a group satisfying the properties in case 2, then there exist a simple abelian variety A, which does not descend to $\mathcal{C}$, and G can be taken equal to $A^{\#}$.

One of the most celebrated application of model theory was to use this to prove function field Mordell–Lang. Here is the characteristic 0 case.

Theorem 1.3.74. *Let $k \subset K$ be algebraically closed subfields of characteristic zero. Let A be an abelian variety defined over K, with k trace zero, i.e. no abelian subvariety of A descends to k.*

Let Γ be a finitely generated subgroup of A, and let X be an irreducible algebraic subvariety of A. Assume $X \cap \Gamma$ is Zariski dense in X. Then X is a coset of an algebraic subgroup $B \subseteq A$.

There is also a positive characteristic version, with a slightly different statement, but which was the really new theorem. Again see [Be98] for more details and references.

1.3.5 *Keisler measures and combinatorics*

We shall discuss the notion of Keisler measures on definable sets and their applications to combinatorics. A basic reference for our point of view is [Pil20] which contains references to original papers. We still use the conventions that T is a complete theory and $\overline{M}$ is κ-saturated, strongly κ-homogeneous for a sufficiently large κ, however we no longer assume T is stable. For convenience, we will occasionally identify a formula $\varphi(\overline{x})$ with the set in $\overline{M}$ it defines, i.e. $\varphi(\overline{M}) = \{\overline{m} \in \overline{M} : \models \varphi(\overline{m})\}$.

Definition 1.3.75. A Keisler measure in $\overline{x}$ (a collection of free variables/sorts) over a set of parameters A is a finitely additive probability measure $\mu(\overline{x})$ on the Boolean algebra of sets defined by A-formulas in variables $\overline{x}$, i.e. if $\varphi(\overline{x})$ defines X, then $0 \leqslant \mu(X) \leqslant 1$ is a real number. Furthermore, $\mu(\overline{M}^{|\overline{x}|}) = 1$, $\mu(\varnothing) = 0$, and if X, Y are definable and $X \cap Y = \varnothing$, then $\mu(X \cup Y) = \mu(X) + \mu(Y)$. We will feel free to abuse notation and write $\mu(\varphi(\overline{x}))$ to mean $\mu(X)$ where $\varphi(\overline{x})$ defines X in $\overline{M}$ (that is $X = \varphi(\overline{M})$.)

Remark 1.3.76.

1. A complete type $p(\overline{x}) \in S_{\overline{x}}(A)$ induces the Keisler measure μ_p, where $\mu(\varphi(\overline{x})) = 1$ if $\varphi(\overline{x}) \in p$ and 0 otherwise. This requires a short proof.
   Because types are finitely consistent, we have $\overline{x} = \overline{x} \in p$, hence $\mu_p(\overline{M}^{|\overline{x}|}) = \mu_p(\overline{x} = \overline{x}) = 1$. Similarly, $\mu_p(\varnothing) = \mu_p(\neg(\overline{x} = \overline{x})) = 0$. Since p is a complete and finitely consistent, then if $\varphi(\overline{M}) \cap \psi(\overline{M}) = \varnothing$, at most one of $\psi(\overline{x})$, $\varphi(\overline{x})$ is in p, so $\mu_p(\varphi(\overline{x}) \vee \psi(\overline{x})) = \mu_p(\varphi(\overline{M}) \cup \psi(\overline{M})) = \mu_p(\varphi(\overline{x})) + \mu_p(\psi(\overline{x}))$.
2. Suppose $\Delta(\overline{x})$ is a finite collection $\{\varphi_1(\overline{x}, \overline{y_1}), \ldots, \varphi_r(\overline{x}, \overline{y_r})\}$ of L-formulas. Then we have the Keisler Δ-measures over A, i.e. where the relevant Boolean algebra is the set of formulas $\psi(\overline{x}) \in L_A$ which are equivalent to Boolean combinations of A-instances of $\varphi_i(\overline{x}, \overline{y_i})$ for $1 \leqslant i \leqslant r$.
3. We shall usually assume A is a model of T, i.e. $A = M$ is an elementary substructure of $\overline{M}$.

Example 1.3.77.

1. Complete types induce Keisler measures as in Remark 1.3.76.
2. Let $T = RCF$. Let $A = M = \mathbb{R}$ be the standard model of T, which is an elementary substructure of $\overline{M}$. Consider μ, the Lebesgue measure on $[0,1]_{\mathbb{R}}$, the standard copy of the unit interval. (Not to be confused with $[0,1]_{\overline{M}}$.)
   Given $\varphi(x)$, let $\varphi'(x)$ be $\varphi(x) \wedge (0 \leqslant x \leqslant 1)$. Then since we have quantifier elimination for real closed ordered fields, φ' is a finite Boolean combination of polynomial equalities and inequalities, so $\varphi'(\overline{M})$ is a finite collection of intervals and points lying in the unit interval.
   Then define $\mu'(\varphi(x)) = \mu(\varphi'(\mathbb{R}))$. Since $\mu([0,1]) = 1$ and $\mu(A) \leqslant \mu(B)$ for $A \subseteq B$, μ' is $[0,1]$-valued, $\mu'(x = x) = \mu((x = x)') = 1$. Since $\mu(\varnothing) = 0$, $\mu'(\varnothing) = 0$ as well. Finite additivity for μ' follows from σ-additivity of μ.
   Remark that the Lebesgue measure can also be used to defined a measure on the sort $\bar{M}^n$, for any n.
3. Let $\{\alpha_i\}_{i\in\omega}$ be such that $\Sigma_{i=0}^{\infty} \alpha_i = 1$. Then for Keisler measures $\{\mu_i(\overline{x})\}_{i\in\omega}$ over A, $\mu(\overline{x}) = \Sigma_{i=0}^{\infty} \alpha_i \mu_i(\overline{x})$ is clearly also a Keisler measure over A. This is called a weighted average of the μ_i's. Of special note is when each of the μ_i's is given by a complete type, then this is called a weighted average of types.

Definition.

- For a topological space X, the Borel σ-algebra of X, denoted $B(X)$ is the collection of sets generated by the open sets under the actions of complementation and taking countable unions, or equivalently, the smallest σ-algebra containing all the open sets of X.
- A Borel probability measure is a σ-additive measure $\mu : B(X) \to [0,1]$.
- A Borel probability measure μ is regular if for any $B \in B(X)$, we have

$$\mu(B) = \inf\{\mu(U) : B \subseteq U \text{ open}\} = \sup\{\mu(C) : C \subseteq B \text{ closed}\}.$$

Note. Note that if X is compact and totally disconnected, i.e. a Stone space, then regularity of a Borel probability measure μ implies that for all closed $B \subseteq X$, $\mu(B) = \inf\{\mu(C) : B \subseteq C \text{ clopen}\}$. That is, μ is determined uniquely by its value on the clopen sets.

Fact 1.3.78. *A Keisler measure $\mu(\overline{x})$ over M an elementary substructure of $\overline{M}$ can be identified with a regular Borel probability measure on $S_{\overline{x}}(M)$. Similarly, a Keisler Δ-measure $\mu(\overline{x})$ over M can be identified with a regular Borel probability measure on $S_\Delta(M)$.*

More specifically, given a Keisler measure $\mu(\overline{x})$ over m, define μ' to be a Borel probability measure μ' on $S_{\overline{x}}(M)$ via $\mu'([\varphi(\overline{x})]) = \mu(\varphi(\overline{x}))$ for all clopen sets $[\varphi(\overline{x})]$. By the above note, this is enough to uniquely determine the value of μ' on all Borel subsets of $S_{\overline{x}}(M)$. (Similarly for Keisler Δ-measures and $S_\Delta(M)$.)

Given a regular Borel probability measure μ' on $S_{\overline{x}}(M)$, define the Keisler measure μ via $\mu(\varphi(\overline{x})) = \mu'([\varphi(\overline{x})])$.

Stability, Cantor–Bendixson rank and Keisler measures come together in the following:

Lemma 1.3.79. *Suppose $\Delta(\overline{x})$ is a finite collection of stable L-formulas (as in Remark 1.3.76(2)). Then any Keisler Δ-measure over M is a weighted average of complete Δ-types.*

Proof. By the above fact, we can identify μ with a regular Borel probability measure on $S_\Delta(M)$. We argue by induction on the maximum rank of elements in the domain of μ. That is, we will prove that for any Stone space of ordinal valued Cantor–Bendixson rank, any Borel probability measure is a weighted average of Dirac measures (i.e. given by types, in the case of a type space). And we will prove this by induction on the maximal Cantor–Bendixson rank of an element (a type, in the case of a type space) of our space.

In the base case, the maximum rank is 0, so all types are isolated, and thus there are only finitely many types, so $\mu(\overline{x})$ can be written as a weighted average of these types. We will now take care of the inductive step.

Note that by Lemma 1.3.34 generalized to finite sets of formulas Δ, the rank $\mathrm{CB}(S_\Delta(\overline{M}))$ is finite. For any partial Δ-type $\Phi(\overline{x})$ over A, recall that:

$$R_\Delta(\Phi(\overline{x})) := \max\{\mathrm{CB}_{S_\Delta(\overline{M})}(p) : \Phi(\overline{x}) \subseteq p(\overline{x})\}.$$

And from Proposition 1.3.36, we have:

$$R_\Delta(\Phi(\overline{x})) = \max\{R_\Delta(p(\overline{x})) : p \in S_\Delta(M) \wedge \Phi(\overline{x}) \subseteq p(\overline{x})\}$$

and there are finitely many types realizing this maximum.

Let $\mu(\overline{x})$ be a Keisler Δ-measure over M. Let $p_1(\overline{x}), \ldots, p_k(\overline{x})$ be the finitely many complete Δ-types over M with $R_\Delta(p_i(\overline{x})) = \mathrm{CB}(S_\Delta(\overline{M}))$. Without loss of generality, assume $\mu(\{p_i(\overline{x})\}) > 0$ for $1 \leqslant i \leqslant r$ for some $0 \leqslant r \leqslant k$ and $\mu(\{p_i(\overline{x})\}) = 0$ for all $r < i \leqslant k$. Define $\alpha_i = \mu(\{p_i(\overline{x})\}) \in (0, 1]$ for $1 \leqslant i \leqslant r$. Let $U = S_\Delta(M)\backslash(\{p_1(\overline{x}), \ldots, p_r(\overline{x})\})$. Then by the additivity of μ, we get

$$\mu(U) = 1 - \sum_{i=1}^{r} \alpha_i.$$

Set $\mu(U) = \beta$. If $\beta = 0$, then notice that

$$\mu(\overline{x}) = \sum_{i=1}^{r} \alpha_i p_i(\overline{x})$$

by finite additivity, where here $p_i(\overline{x})$ stands for the Keisler measure induced by the type $p_i(\overline{x})$ as discussed above.

Therefore, we now assume that $\beta > 0$. By regularity of μ and the above note, for any open set V, we have $\mu(V) = \sup\{\mu(C) : C \subseteq V \text{ clopen}\}$. In particular, $\mu(U)$ is the supremum of a set of reals obtained as the measures of clopen subsets, so we can find a countable sequence of clopen subsets $\varnothing = U_0 \subseteq U_1 \subseteq U_2 \subseteq \cdots \subseteq U$ such that $\lim_{n\to\infty} \mu(U_n) = \beta$, because the reals have a countable dense subset. Set $\beta_i = \mu(U_i)$ for all $i \geqslant 1$.

Each of $U_{i+1}\backslash U_i$ for $i \geqslant 0$ is a clopen, and hence Δ-definable over M, set with positive μ-measure. Furthermore, since all types realizing the maximum rank are not in U and thus not in $U_{i+1}\backslash U_i$ for any $i \geqslant 0$, the rank of the maximum element of each is strictly smaller. Therefore we can apply our induction hypothesis on each to get that

$$\mu(\overline{x}) \upharpoonright (U_{i+1} \backslash U_i) = \sum_{j=1}^{\infty} \beta_{i,j} q_{i,j}(\overline{x})$$

where

$$\sum_{j=1}^{\infty} \beta_{i,j} = \mu(U_{i+1} \backslash U_i).$$

Then by σ-additivity of μ as a regular Borel probability measure, we see that

$$\mu(\overline{x}) = \sum_{i=1}^{r} \alpha_i p_i(\overline{x}) + \sum_{i=0}^{\infty} \sum_{j=1}^{\infty} \beta_{i,j} q_{i,j}(\overline{x}).$$

Noting that

$$\sum_{i=0}^{\infty} \sum_{j=1}^{\infty} \beta_{i,j} + \sum_{i=1}^{r} \alpha_i = 1$$

we conclude that μ is indeed a weighted average of types. □

Remark. In a stable theory, every Keisler measure is locally (formula-by-formula) a weighted average of complete types. However, this is not true of all Keisler measures: In particular, the Keisler measure from Example 1.3.77 given from the standard Lebesgue measure has a unique extension, which cannot be true of a Keisler measure locally equal to a weighted average.

We now discuss an application of Keisler measures to combinatorics. This is the so-called stable regularity lemma (or theorem). It was originally due to Malliaris and Shelah; see [Pil20] for background and references to the original papers. Here we give a treatment making use of "nonstandard" or "pseudofinite" methods and Keisler measures. We shall work in the setting of finite bipartite graphs (L, R, E) where L and R are sets of vertices and $E \subseteq L \times R$ is the edge relation.

Definition. A finite bipartite graph (L, R, E) is said to be ϵ-regular for $\epsilon > 0$ if for any $A \subseteq L$, $B \subseteq R$ and $|A| \geqslant \epsilon |L|$, $|B| \geqslant \epsilon |R|$, we have:

$$\left| \frac{|E \cap (A \times B)|}{|A \times B|} - \frac{|E \cap |L \times R|}{|L \times R} \right| < \epsilon.$$

That is to say, the difference between the density of edges in $(A, B, E \cap (A \times B))$ and the density of edges in the original graph is less than ϵ.

Theorem 1.3.80. *(Szemerédi regularity) For all $\epsilon > 0$, there exists $N_\epsilon \in \omega$ such that for any finite bipartite graph (L, R, E) there are partitions $L = L_1 \cup \cdots \cup L_n$, $R = R_1 \cup \cdots \cup R_m$ for $n, m < N_\epsilon$ and a set of exceptions $\Sigma \subseteq \{1, \ldots, n\} \times \{1, \ldots, m\}$ such that $|(\bigcup_{(i,j)\in\Sigma} L_i \times R_j)| \leqslant \epsilon|L \times R|$ and for every $(i, j) \notin \Sigma$, $(L_i, R_j, E \cap (L_i \times R_j))$ is ϵ-regular.*

Note. From a certain point of view one can see three components in the decomposition:

- The structural component, which is the partition of the graph into substructures of the original.
- The error component, which is the small set of exceptions.
- The pseudo-randomness component, which is the ϵ-regularity asserting that "most" of the graph is regular.

A typical move in combinatorics is to restrict attention to those (finite) graphs omitting (as an induced subgraph) a given finite graph. k-stability means omitting the k-half graph, see below. And under this assumption one obtains a stronger regularity lemma with homogeneity replacing regularity.

Definition. A finite bipartite graph (L, R, E) is said to be ϵ-homogenous if either $|(L \times R)\backslash E| \leqslant \epsilon|L \times R|$ (the graph is "almost" the complete graph) or $|(L \times R) \cap E| \leqslant \epsilon|L \times R|$. (The graph is "almost" the empty graph.)

Definition. The k-half graph (L, R, E) is the graph where $L = R = \{1, \ldots, k\}$ and $E(i, j)$ if and only if $i \leqslant j$.

Theorem 1.3.81. *(Stable Regularity Theorem) For all $k \in \mathbb{N}$ and $\epsilon > 0$, there exists $N_{\epsilon,k} \in \mathbb{N}$ such that whenever (L, R, E) is a finite bipartite graph which omits the k-half graph (i.e. is k-stable), then there are partitions $L = L_1 \cup \cdots \cup L_n$, $R = R_1 \cup \cdots \cup R_m$, $n, m < N_{\epsilon,k}$ such that each of $(L_i, R_j, E \cap (L_i \times R_j))$ is ϵ-homogeneous.*

Note. We can see the conclusion as being purely structural.

We shall prove the Stable Regularity Theorem using a statement about infinite graphs and some pseudofinite methods to apply it to the finite setting.

Theorem 1.3.82. *Let (L, R, E) be a $\varnothing$-definable bipartite graph in the structure M. Assume $E(\overline{x}, \overline{y})$ is stable. Identify E with the formula defining it. Let $\mu(\overline{x})$ be any Keisler measure on L over M. Then for any $1 > \epsilon > 0$ we can find definable partitions $L = L_0 \cup \ldots L_n$, $R = R_1 \cup \ldots R_m$ such that* *for each* $(i,j) \subseteq \{0,\ldots,n\} \times \{1,\ldots,m\}$ *either:*

$$\forall \overline{b} \in R_j \; \mu(L_i \backslash E(\overline{x},\overline{b})) \leqslant \epsilon\mu(L_i)$$

or

$$\forall \overline{b} \in R_j \; \mu(L_i \cap E(\overline{x},\overline{b})) \leqslant \epsilon\mu(L_i).$$

Moreover, each L_i *is defined by an* E*-formula and each* R_j *is defined by an* E^**-formula.*

Proof. Let $\Delta = \{E(\overline{x},\overline{y})\}$ and let μ_0 be the Keisler Δ-measure over M obtained by restricting μ to Δ-formulas. By Lemma 1.3.79, there is I an initial segment of ω (i.e. either finite or infinite) such that $\mu_0 = \Sigma_{i\in I}\,\alpha_i p_i$ where p_i is the measure obtained from a complete Δ-type p_i and $\alpha_i \in (0,1]$ for all $i \in I$. As before, we identify μ_0 with a regular Borel probability measure on $S_\Delta(M)$.

Note that for all $i \in I$, $\mu(\{p_i\}) = \alpha_i$. Indeed, the measure μ_0 is regular, so $\mu_0(\{p_i\}) = \inf\{\mu_0(C) : p_i \in C \text{ clopen}\}$. But for any finite collection of the p_j's, $p_j \neq p_i$, there is a clopen set containing p_i and not containing any of the p_j's. Therefore $\mu_0(\{p_i\}) \leqslant \alpha_i + \zeta$ for any $\zeta > 0$. Lastly, we have $\mu_0(\{p_i\}) \geqslant \alpha_i$ by regularity and the fact that μ_0 is a weighted average of a collection of types containing p_i.

By regularity, for each i there exist a clopen set L_i such that $\mu_0(L_i) < \mu_0(\{p_i\}) + \epsilon\mu(L_i)$ because we can pick L_i with $\mu_0(L_i) < \frac{\alpha_i}{1-\epsilon}$. Then by additivity of μ_0, we get $\mu_0(L_i\backslash\{p_i\}) < \epsilon\mu_0(L_i)$.

Let $B = S_\Delta(M)\backslash\{p_i : i \in I\}$. Then B is Borel as the complement of a countable union of closed sets. Furthermore, since $\mu_0 = \Sigma_{i\in I}\,\alpha_i p_i$, we have $\mu_0(B) = 0$. Now let $\delta = \frac{\alpha_0}{1-\epsilon} - \mu_0(L_0)$. By regularity of μ_0, there is an open set $B \subseteq U$ such that $\mu_0(U) < \delta$.

We have obtained $\{U\} \cup \{L_i : i \in I\}$, an open cover of $S_\Delta(M)$. By compactness there is a finite subcover, which we can assume to be $\{U, L_0, \ldots, L_n\}$ for some n. Furthermore, we can assume these are pairwise disjoint by letting L_i be $L_i\backslash(\bigcup_{j<i} L_{i-1})$ for $1 \leqslant i \leqslant n$. Each L_i is clopen, so $L_1 \cup \cdots \cup L_n$ is clopen, hence its complement is clopen. Let $L_0' = (L_1 \cup \cdots \cup L_n)^c$. Notice that $L_0 \subseteq L_0' \subseteq U \cup L_0$ and $\mu_0(L_0') \leqslant \mu_0(U) + \mu_0(L_0) < \delta + \mu_0(L_0) = \frac{\alpha_0}{1-\epsilon}$, so $\mu_0(L_0') < \frac{\alpha_0}{1-\epsilon}$ as we had before.

Therefore $L_0' \cup L_1 \cup \cdots \cup L_n$ is a definable partition of $L(M)$ with $\mu_0(L_i\backslash\{p_i\}) < \epsilon\mu_0(L_i)$ for all $0 \leqslant i \leqslant n$. For each $0 \leqslant i \leqslant n$, let $\psi_i(\overline{y})$ be the Δ^*-formula defining $p_i \in S_\Delta(M)$, which exists by Proposition 1.3.39 generalized to finite sets of formulas.

For each $J \subseteq \{0, \ldots, n\}$, let $R_J(\overline{y})$ be the formula $\bigwedge_{i \in J} \psi_i(\overline{y}) \wedge \bigwedge_{i \notin J} \neg\psi_i(\overline{y})$. Then the nonempty R_j's are clearly a partition of $R(M)$ by definition. Fix $0 \leqslant i \leqslant n$ and $J \subseteq \{0, \ldots, n\}$ with $R_J(M)$ nonempty. Then if $i \in J$, for any $\overline{b} \in R_J(M)$ we have $E(\overline{x}, \overline{b}) \in p_i$. Hence for any $\overline{b} \in R_j(M)$, the set $[E(\overline{x}, \overline{b})]$ is clopen, containing p_i, and $L_i \backslash [E(M, \overline{b})] \subseteq L_i \backslash \{p_i\}$, so we obtain $\mu_0(L_i \backslash [E(\overline{x}, \overline{b})] < \epsilon\mu_0(L_i)$. If $i \notin J$, then we have $E(\overline{x}, \overline{b}) \notin p_i$. Apply the same argument to get $\mu_0(L_i \cap [E(\overline{x}, \overline{b})]) < \epsilon\mu_0(L_i)$. □

We shall now prove the Stable Regularity Theorem above using some pseudofinite methods.

Proof of Stable Regularity Theorem. Assume not for the sake of a contradiction. Then for some $1 > \epsilon > 0$, no $N \in \omega$ witnesses the existence of partitions with the desired property. Then for each N, there is a counterexample (L_N, R_N, E_N). Hence $|L_N|$ and $|R_N|$ must tend to infinity, as for N large enough the statement is trivial for small enough vertex sets.

Let $\mathcal{L} = \{\in, l, r, e\}$ be the language of set theory together with constant symbols l, r, and e. Let M_n be the $\mathcal{L}$-structure $(\mathbb{V}, L_N, R_N, E_N)$, where $\mathbb{V}$ is the set-theoretic universe. Let Σ be the incomplete $\mathcal{L}$-theory consisting of all sentences satisfied by cofinitely many M_N's. Then Σ is finitely consistent (if we have some finite subcollection, each of them is satisfied by cofinitely many M_N's, and the intersection of finitely many cofinite sets is cofinite), so it is consistent by compactness.

Let $M^* \models \Sigma$ be $(2^{\aleph_0})^+$-saturated, and write $M^* = (\mathbb{V}^*, L^*, R^*, E^*)$. Then (L^*, R^*, E^*) is k-stable since each M_N is k-stable, and this is witnessed by a formula.

Note that each definable subset A of L^* or R^* is assigned a cardinality $|A|$ in M^*, which is a (possibly non-standard) natural number. Indeed, in a model of set theory, cardinality is a definable function, so it lifts to the model M^*.

Hence, for any definable set A in say L^*, we get a nonstandard rational number $\frac{|A|}{|L^*|}$. Now define $\mu(A) := \mathrm{st}(\frac{|A|}{|L^*|})$, where $\mathrm{st} : \mathbb{R}^* \to \mathbb{R}$ is the standard part map, sending each finite non standard real to the unique standard real infinitesimally close to it. This can be seen as a non-standard counting measure, and is indeed a Keisler measure, because the counting measure is.

Therefore we can apply the above theorem to (L^*, R^*, E^*) with μ and $\delta = \frac{\epsilon}{2}$, a real number outside of M^*. (If we apply the theorem in M^*, then the partitions might be indexed by nonstandard naturals, which will not work. One can verify, though, that the proof did not require us to use anything about standard models, so we can apply it outside of the model just as well.) We obtain $L^* = L_1^* \cup \cdots \cup L_n^*$ and $R^* = R_1^* \cup \cdots \cup R_m^*$ for standard naturals n, m, all L_i and R_j definable in M^*, and each $(L_i^*, R_j^*, E^* \restriction (L_i^* \times R_j^*))$ is δ-homogeneous with respect to μ. Then by definition of μ, we have that either for all $\bar{b} \in R_j^*$ $|L_i^* \backslash E^*(\bar{x}, \bar{b})| \leqslant \delta|L_i^*| < \epsilon|L_i^*|$ or for all $\bar{b} \in R_j^*$ $|L_i^* \cap E^*(\bar{x}, \bar{b})| \leqslant \delta|L_i^*| < \epsilon|L_i^*|$.

Notice that there is some sentence σ in $\mathcal{L}$ that expresses this, i.e. uses l to interpret L, r to interpret R, and uses e to interpret E, the language of set theory allows us to express the existence of such a partition by a first order sentence. Thus $M^* \models \sigma$ for infinitely many M_N's, as otherwise $\neg\sigma \in \Sigma$ as it would be true in cofinitely many M_N's. In particular, let $N > n, m$ for which such a partition exists, it is a contradiction, as (M_N, R_N, E_N) was assumed to be a counterexample to the regularity theorem, i.e. assumed to not possess such a partition. □

References

[Bal88] John T. Baldwin. *Fundamentals of Stability Theory.* Perspectives in Mathematical Logic. Springer, 1988.

[Be98] Elisabeth Bouscaren (editor). *Model Theory and Algebraic Geometry: An introduction to E. Hrushovski's proof of the geometric Mordell-Lang conjecture.* Springer, 1998.

[Gro52] A. Grothendieck. Criteres de compacite dans les espaces functionnels generaux. *American Journal of Mathematics*, 74(1):168–186, 1952.

[Hru93] E. Hrushovski. A new strongly minimal set. *Annals of Pure and Applied Logic*, 62(2):147–166, 1993.

[Pil96] Anand Pillay. *Geometric Stability Theory.* Oxford logic guides. Clarendon Press, 1996.

[Pil02] Anand Pillay. Lecture notes - model theory, 2002. `https://www3.nd.edu/~apillay/pdf/lecturenotes_modeltheory.pdf`.

[Pil08] Anand Pillay. *An introduction to stability theory.* Dover Publications, 2008.

[Pil16] A. Pillay. Generic stability and Grothendieck. *South American Journal of Logic*, 2016.

[Pil20] A. Pillay. Domination and regularity. *Bulletin of Symbolic Logic*, 26(2):103–117, 2020.

[PZ03] Anand Pillay and Martin Ziegler. Jet spaces of varieties over differential and difference fields. *Selecta Mathematica*, 9(4):579–599, 2003.

[She90] S. Shelah. *Classification Theory, revised edition.* Studies in Logic. North-Holland, 1990.

[SR94] I. R. Shafarevich and M. Reid. *Basic algebraic geometry*, volume 2. Springer, 1994.

[TZ12] Katrin Tent and Martin Ziegler. *A course in model theory*, volume 40. Cambridge University Press, 2012.

Chapter 2

Continuous Logic

2.1 Introduction

This chapter is based on notes from a course I taught on Continuous Logic at Notre Dame in Spring 2021, which were originally typed up by the graduate students in the course Nicolas Chavarria, Gabriel Day, David Meretzky, Justin Miller, and Gurutam Thockchom. Postdocs Chieu Minh Tran and Rémi Jaoui also sat in on the course and made some helpful contributions. Sometimes the students/scribes were asked to fill in or give details of proofs. As with the chapter on Stability Theory, I have tinkered with the notes quite a bit (including adding some material). In particular the introduction here is new. Nevertheless the material remains rather uneven, even notationwise. I take responsibility for errors.

Continuous logic is about a variant of (finitary) first order logic, where formulas are real-valued rather than $\{0,1\}$-valued. We could also allow formulas to have values in arbitrary compact Hausdorff spaces. A key feature is that the compactness theorem is still valid. We will use the acronym FOL for $\{0,1\}$-valued first order logic, and CL for its continuous variants. The many-valued propositional logic of Lukasiewicz appeared in the early part of the 20th century with philosophical motivation, but I will focus here more on the later mathematical and model-theoretic developments. The formalism for FOL is fairly robust and uncontroversial, although various categorical or topological accounts can also be given. However in the case of CL there are and have been a number of formalisms. An attractive and fairly comprehensive such formalism was developed around the mid 2000's and is expounded in a couple of seminal papers, [BBHU08] and [BU10]. There has been a tendency in recent years to identify the descriptive expression "continuous logic" with the specific technical apparatus and notation described in these papers and the main aim of the course was to give an account of this apparatus. Nevertheless I want to also mention other approaches and also the extent to which one can view CL as already present in FOL via various generalizations of definability.

In Section 2.2 some of these other variants are discussed, including a recent formalism of Chavarria and the author. In Section 2.3 we give an exposition of the [BBHU08] formalism. We follow the terminology of that paper (which is not always optimal or memorable). In Section 2.4 we give an exposition of stability in CL, largely drawn from [BU10], but also touching on recent work of the author with Chavarria Gomez and Conant (whom I thank here).

The reader can see Section 1.2 of the chapter on stability theory (in this volume) for model-theoretic prequisites. There may be a little overlap with the current chapter and some repetitions.

2.2 Background

There exist many precursors to CL. This includes on the model theoretic side Chang and Keisler's 1966 text "Continuous Model Theory" ([CK85]) and on the philosophy side Lukasiewicz's Many Valued Logic. This section will begin with an informal discussion of how some CL phenomena arise within FOL. Secondly, we discuss Henson's logic of positive bounded formulas. Thirdly we discuss hyperdefinability in first order logic. And finally we include a discussion of another recent approach to CL from [CP23].

2.2.1 *Examples from FOL*

Typically we fix a complete theory T in a (possibly many sorted) language L. We will often write $\bar{x}$, $\bar{y}$ etc. for (finite) tuples of variables, although later may just write, x, y, ...

Let $\varphi(\bar{x})$ be an L-formula. One can think of φ as giving a functor from Mod(T), the category of models of T, to the category of sets. Here the morphisms of Mod(T) are elementary embeddings, and the morphisms of Set are inclusion maps. For $M \models T$, $\varphi : M \mapsto \varphi(M) = \{\bar{a} \in M^n \mid M \models \varphi(\bar{a})\}$. Indeed by the Tarski–Vaught test, $M \prec N$ implies $\varphi(M) \subseteq \varphi(N)$. One can then ask: When is such a functor constant?

If for some model M of T $\varphi(M)$ is finite, then $M \models \exists^{=n} x\varphi(x)$ for some n. Since T is complete then the same sentence is true in any model of T, hence $M \prec N$ implies $\varphi(M) = \varphi(N)$, so the functor φ is "constant". On the other hand, if for some model M of T, $\varphi(M)$ is infinite, then by compactness there is an elementary extension $N \succ M$ such that $\varphi(M) \subsetneq \varphi(N)$.

So φ is constant as a functor iff it is finite. In continuous logic, finite will be replaced by compact.

Example 2.2.1. (See also Exercise 2.21 from the Stability Theory chapter.) Take T = RCOF in the language of ordered rings $L = \{+, \times, -, 0, 1, \leqslant\}$. RCOF is $\mathrm{Th}(\mathbb{R}, +, \times, -, 0, 1, \leqslant)$ so it is complete. The theory RCOF can be axiomatized by the axioms for ordered fields together with an axiom schema expressing the intermediate value property.

Let M be a κ-saturated model of RCOF for $\kappa > 2^{\aleph_0}$. Note that $\mathbb{R} \prec M$. Consider $\varphi(x)$ given by $0 \leqslant x \wedge x \leqslant 1$. Let $X = \varphi(M)$. Writing $X = [0, 1](M)$ is only a slight abuse of notation. Further by the unadorned $[0, 1]$ we mean $[0, 1](\mathbb{R})$, equivalently, $\varphi(\mathbb{R})$.

For each $a \in X$ there is a unique $c \in [0, 1](\mathbb{R})$ such that $|a - c| < \frac{1}{n}$ for all $n \in \mathbb{N}$. This follows by the completeness of the real numbers. We say a is "infinitesimally close" to c. The real number c is called the standard part of a. Taking the standard part gives us a map $X \to [0, 1]$, written st and defined by $\mathrm{st}(a) = c$.

Note that the relation $\mathrm{st}(x) = \mathrm{st}(y)$, giving the kernel of the standard part map on X, is defined by infinitely many formulas in the language of ordered rings without parameters:

$$\left\{ |x - y| < \frac{1}{n} \mid n = 1, \ldots \right\}.$$

This collection of formulas defines an equivalence relation $E(x, y)$ for any model M' of $RCOF$. We call E a type-definable equivalence relation.

Via the standard part map, st, $[0, 1]$ and the hyper-definable set $\varphi/E(M)$ are identified. In fact φ/E does not get bigger when passing from M to an elementary extension. Linking this example with the preceding discussion, φ/E is a hyper-definable analogue of a finite definable set.

The next question is: Can one recover the topology on $[0, 1]$, equivalently φ/E just from $\varphi(x)$ and $E(x, y)$? Yes! A subset of φ/E will be closed iff it's preimage is type-definable over a small set of parameters.

Example 2.2.2. Let T be any complete theory (with infinite models). Let $\Sigma(\bar{x})$ be the empty partial type in infinitely many variables $x_1, x_2, \ldots$. Given a model M of T, $\Sigma(M) = M^\omega$. We can view M^ω as a structure in its own right, by for each L-formula φ in variables $x_1, \ldots, x_n$ say, defining $\varphi(M^\omega)$ to be the set of infinite tuples whose first n coordinates satisfy φ (and calling this a basic relation R_φ). Let us assume that the model M of T is saturated. So what kind of structure is M^ω? One can check that if for example elements $\bar{a}$, $\bar{b}$ in M^ω have the same quantifier-free type then there is an automorphism taking one to the other.

On the other hand equality on M^ω is not given by a quantifier-free formula but by a collection of formulas $x_i = y_i$ for $i = 1, 2, \ldots$. This actually gives a metric d on M^ω where $d(\bar{a}, \bar{b}) = 1/n$ where n is greatest such that $(a_1, \ldots, a_n) = (b_1, \ldots, b_n)$ (or 0 if the infinite tuples are identical).

Example 2.2.3. We now consider *normed vector spaces.* Consider a vector space V over the (ordered) field $\mathbb{R}$. A *norm* on V is a real-valued function $||\cdot|| \colon V \to \mathbb{R}$ satisfying the following properties: for all $x, y \in V$ and $\alpha \in \mathbb{R}$,

1. $||x|| \geqslant 0$;
2. $||x|| = 0$ if and only if $x = 0$;
3. $||x + y|| \leqslant ||x|| + ||y||$; and
4. $||\alpha x|| = |\alpha| ||x||$.

Notice that any such norm induces a metric $d(x, y)$ on V defined by

$$d(x, y) = ||x - y||.$$

When the normed vector space V is finite dimensional, it is essentially $\mathbb{R}^n$ equipped with the standard norm $||(x_1, \ldots, x_n)|| = \sqrt{x_1^2 + \cdots + x_n^2}$.

The situation becomes more interesting for infinite dimensional vector spaces. The typical examples of such spaces are spaces of functions, the objects of study in functional analysis.

One example of a function space is $C[a, b]$, the space of continuous functions on an interval (where $a, b \in \mathbb{R}$ and $a < b$). The set of such functions is a vector space under pointwise addition and scalar multiplication, i.e. we define $(f + g)(x) = f(x) + g(x)$ and $(\alpha f)(x) = \alpha f(x)$.

Unlike the finite dimensional spaces considered above, $C[a, b]$ has several commonly-studied norms. These include:

- The L^1 norm: $||f||_1 = \int_a^b |f(x)| dx$;
- The L^2 norm: $||f||_2 = (\int_a^b |f(x)|^2 dx)^{\frac{1}{2}}$; and
- The L^∞ norm: $||f||_\infty = \max\{f(x) \colon a \leqslant x \leqslant b\}$.

To see that $C[a, b]$ is an infinite dimensional space, note that the polynomial functions $1, x, x^2, \ldots$ are an infinite collection of linearly independent continuous functions on any interval $[a, b] \subseteq \mathbb{R}$.

Now we consider a normed vector space as a first order structure $\mathcal{V}$. This is two sorted, with a sort for the vector space and a sort for the real numbers. It contains a norm function $||\cdot||$ as well as scalar multiplication functions $\lambda_r \colon V \to V$ for each real number r:

$$\mathcal{V} = \langle (V; +, \cdot, 0), (\mathbb{R}; +, \cdot, -, 0, 1, <), || \cdot ||, \lambda_r \text{ for each } r \in \mathbb{R} \rangle.$$

We now consider an elementary extension $\mathcal{V}^*$ of $\mathcal{V}$ which is κ-saturated for some $\kappa > |\mathcal{V}|$. This new structure has the same form as $\mathcal{V}$, with κ-saturated elementary extensions V^* and $\mathbb{R}^*$ of V and $\mathbb{R}$ respectively, and a new "norm" function $|| \cdot ||^* \colon V^* \to \mathbb{R}^*$. Here V^* is an "$\mathbb{R}^*$ normed vector space", no longer a vector space over $\mathbb{R}$.

Define

$$V^*_{fin} = \{x \in V^* \colon ||x||^* \text{ is finite}\},$$

where "finite" here means $\leqslant n \in \mathbb{R} < \mathbb{R}^*$, for some natural number n. Now, define an equivalence relation E on V^*_{fin} much as we did on models of RCOF: $E(x, y)$ holds if and only if $||x - y||^* <^* \frac{1}{n}$ for all $n \in \mathbb{N}$.[1]

Then V^*_{fin}/E has naturally the structure of a normed vector space over $\mathbb{R}$. This of course has to be checked. For example; Let $\pi \colon V^*_{fin} \to V^*_{fin}/E$ be the map sending an element of V^*_{fin} to its equivalence class modulo E. We show that for $v_1, v_2 \in V^*_{fin}$, $\pi(v_1 + v_2) = \pi(v_1) + \pi(v_2)$. By definition $\pi(v_1) = \{u \in V^*_{fin} \colon ||v_1 - u||^* < \frac{1}{n}, \text{ for all } n\}$, and similarly for $\pi(v_2)$ and $\pi(v_1 + v_2)$. Suppose that $u_1 \in \pi(v_1)$ and $u_2 \in \pi(v_2)$. Then

$$||v_1 + v_2 - (u_1 + u_2)||^* = ||(v_1 - u_1) + (v_2 - u_2)||^* \leqslant ||v_1 - u_1||^* + ||v_2 - u_2||^*,$$

where inequality comes from the triangle inequality on $|| \cdot ||^*$. Since $u_1 \in \pi(v_1)$ and $u_2 \in \pi(v_2)$, we have that both $||v_1 - u_1||^*$ and $||v_2 - u_2||^*$ are both smaller than $\frac{1}{n}$ for any n, so the same holds for their sum. Thus $u_1 + u_2 \in \pi(v_1 + v_2)$. By an identical triangle inequality argument, we can get that for $u \in \pi(v_1 + v_2)$ and $u_1 \in \pi(v_1)$, we have $u - u_1 \in \pi(v_2)$, and reciprocally $u - u_2 \in \pi(v_1)$. We conclude that $\pi(v_1) + \pi(v_2) = \pi(v_1 + v_2)$ as desired.

Likewise we define scalar multiplication by showing that for r real, and $x, y \in V^*_{fin}$, $E(x, y)$ iff $E(rx, ry)$. And define the norm of the E-class of x to be the standard part of $||x||^*$, show it is well defined.

Let

$$B_n = \{x \in V^* \colon ||x||^* \leqslant n\}.$$

The restriction of E to each of these B_n is type-definable over $\varnothing$ by the formulas $\{||x - y||^* < \frac{1}{n} \colon n \in \mathbb{N}\}$. So each B_n/E is a "hyperdefinable" set, and V^*_{fin}/E is the (countable) union.

[1] From now on we will abuse notation slightly by referring to the ordering $<^*$ in the saturated structure by $<$.

For various reasons, there is a kind of preference among many circles for ultrapower (and ultraproduct) constructions, maybe because they are considered more explicit, or even "algebraic". And the above construction of V^*_{fin}/E from V was carried out in [DCK71] by such methods. Moreover these normed space ultrapowers (and ultraproducts) became standard tools in functional analysis.

Now Łoś's Theorem says that first order formulas are preserved by usual ultrapowers and also by ultraproducts when suitably formulated. So it is natural for a logician to ask what kind of "formulas" are preserved by the normed space ultrapower construction (taking V to V^*_{fin}/E). We understand this to be one of the motivations for Ward Henson to develop his positive bounded logic (described below).

Example 2.2.4. Here we give an informal account of a "CL-theory" attached to a classical first order theory T. It is partly drawn from Section 3.1 of [HKP22] and overlaps with parts of the Stability Theory chapter of this volume.

We fix a complete classical first order theory T in language L, one-sorted for simplicity. As usual, $\bar{M}$ is a very saturated model of T, M, N (small) elementary substructures of $\bar{M}$ and $A, B, ...$ (small) subsets of $\bar{M}$. We have the type spaces $S_n(A)$ and even $S_\omega(A)$.

Let C be a (Hausdorff) topological space. By a C-valued formula in n-variables $x_1, ..., x_n$ over A, we mean a continuous function φ from $S_n(A)$ to C. (When $C = \{0,1\}$ a C-valued formula over A corresponds to an L-formula over A.) Note that the image of φ is compact so we may assume C to be compact. Such a formula defines a map $\varphi(M)$ from M^n to C for each model M containing A (and also for the monster model $\bar{M}$), by $\varphi(M)(\bar{b}) = \varphi(tp(\bar{b}/A))$. As at the beginning of Section 2.2.1, the formula φ gives a functor which takes $M \to \varphi(M)$, which we call an A-definable function from n-tuples to C. All this makes sense for ω-tuples too. We identify the C-valued formula φ over A with $\varphi(\bar{M})$.

By a CL-formula over A we mean a $\mathbb{R}$-valued formula over A.

Let us give a summary without proofs of some basic (local) stability results for CL-formulas generalising material in Section 1.3 of the Stability Theory chapter. These will appear again in fully fledged continuous logic in Section 2.4.

First we let $\varphi(x, y)$ be a CL-formula of T, over $\emptyset$, where x, y are tuples of variables (which will here take to be finite tuples, although ω-tuples would be OK). We define φ to be stable (for T) if for each $\epsilon > 0$ there do not exist a_i, b_i from $\bar{M}$ for $i < \omega$ such that $|\varphi(a_i, b_j) - \varphi(a_j, b_i)| \geqslant \epsilon$ for all $i < j$. It is an exercise that the CL-formula $\varphi(x, y)$ is stable for T if and only if whenever $(a_i, b_i)_{i<\omega}$ is indiscernible then $\varphi(a_i, b_j) = \varphi(a_j, b_i)$ for all $i < j$.

If T is stable as a (classical) first order theory, then every CL-formula will be stable.

Given a tuple a from $\bar{M}$ of the same length as x, by the complete φ-type of a over a model M we mean the function taking a tuple b from M to $\varphi(a, b)$. We denote this $tp_\varphi(a/M)$. We will call this type *definable* iff it is a definable over M function on tuples of length that of y.

Proposition 2.2.5. *Let $\varphi(x, y)$ be a CL-formula of T. Then the following are equivalent:*

(i) φ is stable for T,
(ii) Every complete φ-type over a model is definable.

It turns out that formulas with values in compact spaces, as we have defined here, have a natural correspondence with another model-theoretic object, *type-definable bounded equivalence relations.*

Definition 2.2.6. Let $\Sigma(\bar{x}, \bar{y})$ be a (partial) type consistent with T (where the barred variables are all n tuples for a single n). If, for all models $M \models T$, we have that the interpretation $\Sigma(M)$ of Σ in M is an equivalence relation on n-tuples, then this equivalence relation is said to be *type-definable.* If there is a cardinal κ such that for all models M of T, $|M^n/\Sigma(M)| \leqslant \kappa$, then this equivalence relation is *bounded.*

As suggested above, formulas taking values in a compact space C correspond to bounded type-definable equivalence relations via a natural mapping. That is to say, given a C-valued formula one can define a type-definable equivalence relation which is bounded, and given a bounded type-definable equivalence relation one may recover a C-valued formula for appropriate C.

We sketch one direction of this correspondence. Given C-valued $\varphi\colon S_n(T) \to C$, we define E. Let $\bar{M}$ be a $|C|^+$ saturated model of T, such that all types in $S_n(T)$ are realized. As discussed above when defining C-valued relations, we get a map $\varphi(\bar{M})\colon \bar{M}^n \to C$ mapping a tuple $\bar{a}$ to $\varphi(tp(\bar{a}))$. Thus we get an equivalence relation $E(\bar{x}, \bar{y})$ on n-tuples given by $E(\bar{x}, \bar{y})$ if and only if $\varphi(\bar{M})(\bar{x}) = \varphi(\bar{M})(\bar{y})$. Let us briefly check type-definability and boundedness of E. First the set of (p, q) in $S_n(T) \times S_n(T)$

such that $\varphi(p) = \varphi(q)$ is closed in the product topology so also in the space $S_{2n}(T)$, so Σ is type-definable (without parameters). By the saturation assumption on $\bar{M}$, E is bounded.

2.2.2 *Henson's positive bounded logic*

The basic reference to this section is [HIKO03]. As mentioned earlier the motivation for developing this logic was to characterise the "formulas" preserved by the Banach space ultraproduct and ultrapower.

A *Banach space* is a normed vector space which is *complete*, that is to say, such that all Cauchy sequences in the space converge. We assume here that all Banach spaces are over $\mathbb{R}$. Examples of Banach spaces include the space of continuous functions under the norm

$$||f||_p = \left(\int_a^b |f|^p\right)^{\frac{1}{p}}.$$

They also include finite dimensional spaces $\mathbb{R}^n$ with the usual norm. Also the l_p sequence spaces consisting of the suitable sequences $(x_1, x_2, ...)$ of reals with the norm

$$||\bar{x}|| = \sqrt[p]{|x_1|^p + |x_2|^p + \cdots}.$$

We will start with one of motivating problems and results.

Example 2.2.7. Let T be a linear map between normed vector spaces V, W. Then a norm can be defined for T by

$$||T|| = \sup\{||Tv||: ||v|| \leqslant 1\}.$$

By a λ-isomorphism between V and W we mean an isomorphism T of vector spaces such that both $||T||$ and $||T^{-1}|| \leqslant \lambda$. Here $\lambda \geqslant 1$, and note that when $\lambda = 1$, T is an isometry, namely a norm preserving isomorphism.

The following kind of problem was mentioned in [DCK71]:
Given a class $\mathcal{C}$ of Banach spaces and a Banach space B, suppose that every finite dimensional subspace of B is λ-isomorphic to some $C \in \mathcal{C}$, under what conditions is B itself λ-isomorphic to some $C \in \mathcal{C}$?

It was shown in [DCK71] that the problem has a positive answer (for all $\lambda \geqslant 1$) when $\mathcal{C}$ is a class of Banach spaces closed under subspace, isomorphism, and Banach space ultraproducts.

The point of this discussion above is that closure under Banach space ultraproducts can be replaced by closure under a suitable notion of elementary equivalence, with respect to suitable notions of formula, sentence etc. What is the appropriate logic, and can it be generalized?

In [HIKO03] this is put in the context of normed space structures, where there are some additional functions. This includes the case of Banach space lattices.

We give an exposition of the logic. In [HIKO03] many-sorted normed space structures are considered, but for simplicity we just consider the 1-sorted case.

We will first describe a suitable structure, then the appropriate language or vocabulary for such a structures, then the formulas.

We have two distinguished sorts V and $\mathbb{R}$ where V is an $\mathbb{R}$ vector space. We have:

- $0, +, -$ V;
- $+, \times, -, 0, 1$ on $\mathbb{R}$;
- Constant symbols $q\colon \mathbb{R} \to \mathbb{R}$ for each $q \in \mathbb{Q}$;
- Norms $||\cdot||\colon V \to \mathbb{R}$ and $|\cdot|\colon \mathbb{R} \to \mathbb{R}$ (note that the former restricts to the latter on $\mathbb{R}$);
- Scalar multiplication $\lambda_r\colon V \to V$ for each $r \in \mathbb{R}$;
- Certain distinguished functions from products of sorts to sorts, such as $V \times V \to V$, $V \times \mathbb{R} \to \mathbb{R}$, and so on. These functions are required to be uniformly continuous on bounded sets (with respect to the norms).

For the syntax we give ourselves an appropriate signature or language with symbols for the data above. But this will be a 2-sorted functional language. Moreover the second sort is fixed as $\mathbb{R}$. But the first sort as well its norm can vary.

We now define formulas inductively, as in first order logic. It is important to note here that our use of the symbols $\leqslant, \geqslant$ is entirely *formal*; these are *not* relation symbols in the language for a structure.

- **Terms** are defined to be: variables and functions applied to other terms.
- **Atomic formulas** are $q \leqslant t$ and $t \leqslant q$, where t is a real-valued term or variable of the real sort, and $q \in \mathbb{Q}$ (or $q \in \mathbb{R}$ if we allow constant symbols for elements of $\mathbb{R}$).
- **Formulas** include precisely:
  1. Atomic formulas;
  2. Closure under connectives $\varphi \wedge \psi$ and $\varphi \vee \psi$, where φ and ψ are formulas;
  3. Closure under quantifiers $\exists x(||x|| \leqslant q \wedge \varphi)$ and $\forall x(||x|| \leqslant q \to$

φ) for $q \in \mathbb{Q}$ and formula φ. We sometimes abbreviate these formulas as $\exists_q x\varphi$ and $\forall_q x\varphi$.

The formulas above are called *positive bounded (pb) formulas.*

We have the obvious semantics, and for M a structure, write $M \models \varphi(\bar{a})$ if and only if φ is true of $\bar{a}$ in M.

A sentence is a formula without free variables. We now give an example to show how to translate ordinary sentences into this syntax, and also discuss the semantic conditions under which the given sentence holds.

Example 2.2.8. We can write down a pb sentence σ expressing that there are v and w in the V-sort such that $v, w, v + w$, and $v - w$ all have norm 1. For example,

$$\sigma = \exists_1 v \exists_1 w(||v|| \geqslant 1 \wedge ||w|| \geqslant 1 \wedge ||v + w|| \leqslant 1 \wedge ||v + w|| \geqslant 1 \wedge ||v - w|| \leqslant 1 \wedge ||v - w|| \geqslant 1).$$

Note that we have assumed that v, w are variables ranging over the V sort. Also, in the future we can write $t = q$ as an abbreviation for $t \leqslant q \wedge t \geqslant q$.

Under what conditions on V does σ hold? In fact σ holds if and only if V contains a copy of $\mathbb{R}^2$ under the norm $||(x, y)|| = \max\{|x|, |y|\}$.

To see that this is the case, first note that if V contains a copy of $\mathbb{R}^2$ with this norm, $(1, 0)$ and $(0, 1)$ satisfy the sentence above. Conversely, if v, w satisfy the sentence, then they will be linearly independent, so generate a copy of $\mathbb{R}^2$. One can check that the induced norm is as stated.

Let us discuss approximations of positive bounded formulas. These approximations are obtained by *weakening* all the inequalities in any given formula. If φ' is an approximation of the pb-formula φ, we write either $\varphi < \varphi'$ or (equivalently) $\varphi' > \varphi$. More formally, we define *approximations* by induction, as follows:

(i) The approximations of $t \leqslant r$ and $r \leqslant t$ are, respectively, $t \leqslant r'$ for any $r' > r$, and $r' \leqslant t$ for any $r' < r$.

(ii) The approximations of $(\varphi \wedge \psi)$ are $(\varphi' \wedge \psi')$, where φ' and ψ' are approximations of φ and ψ, respectively. We define the approximations of $(\varphi \vee \psi)$ in a similar manner.

(iii) The approximations of $\exists_r x\varphi$ are $\exists_{r'} x\varphi'$, where $r' > r$ and $\varphi' > \varphi$. Similarly, the approximations of $\forall_r x\varphi$ are $\forall_{r'} x\varphi'$, where $r' < r$ and $\varphi' > \varphi$.

Note that φ and its approximations have the same free variables. Moreover, for all models M and all tuples $\overline{a}$ from M, $M \models \varphi(\overline{a})$ implies $M \models \varphi'(a)$ for all approximations $\varphi' > \varphi$.

Next, we define approximate satisfaction of formulas: we say that M *approximately satisfies* φ at $\overline{a}$, and write $M \models_{\mathcal{A}} \varphi(\overline{a})$, if $M \models \varphi'(\overline{a})$ for all $\varphi' > \varphi$. We also have a notion of structures being *approximately elementarily equivalent*: we write $M \equiv_{\mathcal{A}} N$ if, for all pb-sentences σ, $M \models_{\mathcal{A}} \sigma$ if and only if $N \models_{\mathcal{A}} \sigma$. Likewise with the notion of *approximate elementary substructures* and *extensions*: $M \preceq_{\mathcal{A}} N$ if, for any pb-formula $\varphi(\overline{x})$ and any tuple $\overline{a}$ from M, $M \models_{\mathcal{A}} \varphi(\overline{a})$ if and only if $N \models_{\mathcal{A}} \varphi(\overline{a})$.

This notions were applied in Corollary 7.2 of [HIKO03]:

Proposition 2.2.9. *Suppose X, Y are normed vector spaces (considered as structures in the "minimal language" required for such spaces), and suppose $X \preceq_{\mathcal{A}} Y$. Then, for any finite-dimensional subspace E of Y, and any $\eta > 1$, there exists an η-isomorphism T between E and a subspace of X such that T is the identity on $E \cap X$.*

See Chapter 7 of [HIKO03] for the elementary proof. As mentioned there, this result was proved by Stern [Ste76] and by Henson and Moore [HLCM74] [HM74], except with $X \preceq_{\mathcal{A}} Y$ replaced by "Y is a 'Banach space ultrapower of X'".

Next, we consider the Compactness Theorem for positive bounded logic. This is Proposition 9.22 in [HIKO03], although there it is phrased in the language of ultraproducts. Let us fix a signature L for enriched normed vector spaces. Then we have the following:

Proposition 2.2.10 (Compactness Theorem). *Let Γ be a set of pb sentences in L. Suppose its set of approximations Γ^+ is finitely satisfiable, i.e. for every finite subset $\Sigma \subseteq \Gamma^+$, there is an L structure M such that $M \models \Sigma$. Then there is an L-structure M such that $M \models \Gamma$ (so also $M \models \Gamma^+$).*

(There is a caveat regarding some degree of uniformity in the interpretations of the additional function symbols of L that is required for the proof to go through. See Chapter 8 of [HIKO03].)

Proof. We give a sketch proof, using the compactness theorem for first order logic together with the standard part map (rather than ultraproducts as in [HIKO03]).

We view our structures as usual 2-sorted classical first order logic structures in the same vocabulary L. There is no harm in adding constant symbols for elements of $\mathbb{R}$. The pb formulas can be viewed as ordinary L-formulas.

Now, Γ^+, viewed as a set of first order L-sentences, is also finitely satisfiable. So, by the usual compactness theorem, we can choose a κ-saturated model $M^* = (V^*, R^*, \ldots)$ of Γ^+, for some large κ.

As discussed earlier, we let V be V^*_{fin}/E, where V^*_{fin} is the set of finite elements of V^* and E is the equivalence relation $\|x - y\|$ is infinitesimal. Likewise we have $\mathbb{R}^*_{fin}/E$ which is precisely $\mathbb{R}$. The uniform boundedness (and continuity) assumptions on the additional function symbols in L allow us to consider $M = (V, \mathbb{R}, ...)$ as a structure as defined earlier.

We want to show that $M = (V, \mathbb{R}, ...)$ is a model of the set Γ of positive bounded sentences (not just an approximate model).

As an example we consider the pb sentence σ: $\exists x(\|x\| \leqslant r_1 \wedge F(x) \leqslant r_2)$ which we assume to be in Γ. The approximations to σ are of the form σ': $\exists x(\|x\| \leqslant r_1' \wedge F(x) \leqslant r_2')$ for some reals $r_1' > r_1$ and $r_2' > r_2$. Let $\Sigma(x)$ be the set of these approximations as first order formulas. As Σ is finitely satisfiable in M^*, it is realized in M^*, by saturation of M^*. Let a be such a realization. So $a \in V^*_{fin}$ and note $F(a) \in \mathbb{R}^*_{fin}$.

Let π denote the "standard part maps" from V^*_{fin} to V and also from $\mathbb{R}^*_{fin}$ to $\mathbb{R}$. Then we see that as a is $\leqslant$ every real $< r_1$, and $F(a)$ is $\leqslant$ every real $< r_2$, $\pi(a)$ better be $\leqslant r_1$ and $\pi(F(a)) \leqslant r_2$. So σ is true in $M = (V, \mathbb{R}, ...)$.

Of course for arbitrary positive bounded sentences there is an inductive argument involving formulas with free variables and quantifiers. □

A possible "take-away" from the above proof is that positive bounded logic is the logic of the standard part map.

In spite of the subsequent developments in continuous logic, this exposition [HIKO03] is valuable and worth reading. It includes a discussion of quantifier elimination (in the approximate setting) as well as the axiomatizability of various classes of structures by a positive bounded theory. For example the class of Banach lattices isomorphic to $L_p(\mu)$ for some atomless measure μ is pb axiomatizable, and the corresponding theory has QE.

2.2.3 *Hyperdefinability in first order logic*

Fix a complete FOL theory T in a (1-sorted) language L. We have already discussed "imaginaries". In other words, suppose we have an L-formula $\psi(\overline{x}_1, \overline{x}_2)$ (without parameters) such that T "says" $\psi(\overline{x}_1, \overline{x}_2)$ defines an equivalence relation on n-tuples. Fix $M \models T$ and $\bar{a} \in M^n$. Let E be the equivalence relation defined by $\psi(\bar{x}_1, \bar{x}_2)$. We call $\bar{a}/E$, the equivalence class of $\bar{a}$, an *imaginary element.* Passing from T to T^{eq} allows us to consider such $\bar{a}/E$ as elements of a new sort $s_E(M)$. In particular, there is no problem in talking about $\varphi(\bar{x})/\psi(\bar{x}_1, \bar{x}_2)$ as a structure in its own right.

This is totally natural. Many important objects of model theory, such as canonical bases of stationary types in stable theories, may only exist as imaginary elements. In fact adjoining imaginaries is a bit like trying to solve the equation $x^2 = -1$ by introducing imaginary numbers.

What about hyperimaginaries? Let $\Sigma(\bar{x})$ be a partial type (say without parameters). Let $\Gamma(\bar{x}_1, \bar{x}_2)$ be a *set* of L-formulas such that, in any model $M \models T$, Γ defines an equivalence relation on the set $\Sigma(M)$ of realizations of Σ in M.

For $M \models T$, $\bar{a} \in M^n$, and $E = \Gamma(M)$, we call $\bar{a}/E$ a *hyperimaginary.* (Of course, we can also extend this to the case where $\bar{x}$ is an ω-tuple of variables.) Note, however, that in this case, in contraposition to the case of imaginaries, we cannot in general view M^n/E as a first order structure (with the induced structure).

Often we can take Σ to be just $\bar{x} = \bar{x}$. To "see" hyperimaginaries one should work in a saturated model. Σ/Γ is called a *hyperdefinable* set.

The systematic introduction of hyperimaginaries into first order logic occurred in the mid nineties in connection with the study of simple theories. Hart, Kim and I wanted to define *canonical bases* of types in simple theories (in analogy with the stable case). But as far as we know these only exist as hyperimaginaries. Adjoining new sorts for hyperimaginaries to get another first order theory does not really work. Ben Yaacov's compact abstract theories were introduced partly as to provide a formalism for adjoining such sorts (as far as I understand). But in the earlier [HKP00], we gave an ad hoc account of how to work with hypermaginaries (types, indiscernibles, forking, ...) without any new logic formalism. We discuss some of this next.

So, we have a 1-sorted L-theory T and $\Gamma(\bar{x}, \bar{y})$ a "type-definable" equivalence relation on n-tuples (i.e. $\Gamma(M)$ is an equivalence relation on M^n for all $M \models T$), which we call E. Let $\bar{a} \in M^n$ and $B \subseteq M$. What could we mean by $\mathrm{tp}_M((\bar{a}/E)/B)$? Well, let $\varphi(\bar{x}, \bar{b})$ be a formula with parameters from B. Let $\varphi(\bar{x}/E, \bar{b})$ denote the expression "$\exists \bar{y}(\Gamma(\bar{x}, \bar{y}) \wedge \varphi(\bar{y}, \bar{b}))$".

But this is an infinitary formula. However we can consider the set of "approximations" to this expression. Let us assume that $\Gamma(\bar{x}, \bar{y}) = \{\psi_i(\bar{x}, \bar{y}) : i \in I\}$ and that this set is closed under finite conjunctions. Let $\varphi^+(\bar{x}/E, \bar{b})$ be the partial type $\{\exists y(\psi_i(\bar{x}, \bar{y}) \wedge \varphi(\bar{y}, \bar{b})) : i \in I\}$. Then we define $\mathrm{tp}_M((\overline{a}/E)/B)$ to be the union of the partial types $\varphi^+(\overline{x}/E, \overline{b})$ which are true of $\overline{a}/E$.

Note, if $\overline{a}' \in M^n$ and $E(\overline{a}, \overline{a}')$, then $\mathrm{tp}_M((\overline{a}/E)/B) = \mathrm{tp}_M((\overline{a}'/E)/B)$. Similarly, if $M \preceq N$, then $\mathrm{tp}_M((\overline{a}/E)/B) = \mathrm{tp}_N((\overline{a}/E)/B)$.

The main point here is the following. Assume M is $|L^+|$-saturated. Then in M, $\varphi^+(\overline{x}/E, \overline{b})$ is equivalent to "$\exists \overline{y}(\Gamma(\overline{x}, \overline{y}) \wedge \varphi(\overline{y}, \overline{b}))$". The right-to-left implication is obvious. For the left-to-right implication, we take $\overline{a} \in M$ satisfying $\varphi^+(\overline{x}, \overline{b})$ and show that $\{\psi_i(\overline{a}, \overline{y}) \wedge \varphi(\overline{y}, \overline{b}) : i \in I\}$ is consistent, so realized in M.

This is how we describe $\mathrm{tp}_M(e/B)$ where e is a hyperimaginary $\bar{a}/E$ and B is a subset of M. We can also talk about $\mathrm{tp}_M(e/B)$ where e is a hyperimaginary and B is a set of hyperimaginaries. For this see [HKP00].

Remember that $\bar{M}$ is a "monster model" of T, (i.e. κ-saturated and strongly κ-homogeneous for some large κ). The group $\mathrm{Aut}(\bar{M})$ acts on imaginaries and hyperimaginaries. Specifically, if $e = \bar{a}/E$ is a hyperimaginary then so is $f(e) = f(\bar{a})/E$. Moreover, we have that in $\bar{M}$, $\mathrm{tp}(e_1/B) = \mathrm{tp}(e_2/B)$, where e_1 and e_2 are hyperimaginaries and B a small ($< \kappa$) set of hyperimaginaries, if and only if there is $f \in \mathrm{Aut}(\bar{M})$ such that $f(e_1) = e_2$ and f fixes B pointwise. That is, strong κ-homogeneity extends to hyperimaginaries.

We can also talk about indiscernible sequences of hyperimaginaries $(\bar{a}_1/E, \bar{a}_2/E, \bar{a}_3/E, ...)$, meaning that

$$\mathrm{tp}(\bar{a}_{i_1}/E, \ldots, \bar{a}_{i_n}/E) = \mathrm{tp}(\bar{a}_{j_1}/E, \ldots, \bar{a}_{j_n}/E)$$

whenever $i_1 < \cdots < i_n$ and $j_1 < \cdots < j_n$. Thus we can talk about dividing, forking, NIP, etc. in the hyperimaginary context. This is discussed in [HKP00].

Finally, we discuss stability and elimination of hyperimaginaries.

Recall that for T a complete first order theory a formula $\varphi(x, y)$ (where x, y are finite tuples of variables) is stable if there do not exist a_i, b_i for $i < \omega$ in any model M such that $M \models \varphi(a_i, b_j)$ iff $i \leqslant j$.

We can also talk about a sort S or even a formula $\chi(x)$ or partial type $\Sigma(x)$ being stable with respect to the ambient theory T. For example $\Sigma(x)$ being stable means there exists no formula $\varphi(x, y)$ and $M \models T$ and a_i, b_i in M for $i < \omega$ such that each a_i realizes Σ and again $M \models \varphi(a_i, b_j)$ iff $i \leqslant j$.

Note that here the b_i are not required to be tuples of realizations of Σ.

It also makes sense to speak of the stability of an imaginary sort $S_E = M^n/E$ by passing to T^{eq}.

An example is (any completion of) the theory ACVF of algebraically closed valued fields, where the residue field is a stable sort.

Here is an appropriate definition of stability for a hyperimaginary sort $\bar{M}^n/E$: there do NOT exist a_i, b_i for $i < \omega$, such that $a_i \in \bar{M}^n/E$, b_i are tuples from $\bar{M}$, $((a_i, b_i) : i < \omega)$ is an indiscernible sequence, and $tp(a_i, b_j) \neq tp(a_j, b_i)$ for $i < j$.

This can be checked to coincide with the earlier definition for an imaginary sort. There exists first order theories where every imaginary sort is unstable, but there exist stable hyperimaginary sorts.

See Section 1.2 of the chapter on Stability Theory for elimination of imaginaries (EI) for a complete first order theory T. There is a corresponding notion "elimination of hyperimaginaries":

Definition 2.2.11 (Elimination of Hyperimaginaries). T is said to have EHI (elimination of hyperimaginaries) if for any type-definable (over $\varnothing$) equivalence relation E on $\bar{M}^n$) (or even on $\bar{M}^\omega$) and tuple $\bar{a}$, there is a possibly infinite tuple $\bar{b}$ of elements from $\bar{M}^{eq}$ such that $\bar{a}/E$ and $\bar{b}$ are interdefinable over $\varnothing$ (i.e. an automorphism fixes $\bar{a}/E$ iff it fixes $\bar{b}$).

An equivalent condition is that given $\bar{a}$ and E as above, and letting $p(\bar{x}) = tp(\bar{a})$, there is a family of $\varnothing$-definable equivalence relations E_i for $i \in I$ such that on realizations of p, E is equivalent to the conjunction of the E_i.

It is open whether simple theories have EHI. On the other hand stable theories and supersimple theories do have EHI. See [Cas11] for references and a detailed account of some of these results.

2.2.4 *First order structures with a map to a compact space*

Here I briefly and informally describe an account of continuous logic from [CP23], which was NOT part of my original 2021 course. The reader is invited to look at [CP23] for more details.

It is close in spirit to the positive bounded logic from Section 2.2.2. The motivation of the paper was to generalise the notion of an *abelian structure* by adding a homomorphism from the underlying abelian group A to a compact Hausdorff group (such as the circle group), and prove stability. In fact we adapted so-called *pp*-elimination to this context. Anyway it was convenient not to work in the official continuous logic of Chapter 3, but in some kind of amalgam. The idea is to take a first order structure and adjoin to it a C-valued formula where C is a compact space. We saw already in

Section 2.2.1 the notion of a C-valued formula for a first order theory T, but here we are adding such a formula which may not be present in the original theory T.

The most general situation would be a many sorted first order structure equipped with a collection of maps from various sorts to various compact Hausdorff spaces. For simplicity of notation and presentation we just consider a one-sorted first order structure M in language L, together with a map f from the universe of M to a compact Hausdorff space C. We will also assume that $f(M)$ is *dense* in C.

We call this triple (M, f, C) a (C, L)-structure.

What about language and logic? We take a 2-sorted language with a sort of M and a sort for C, with all the L-structure in the first sort, with a function symbol for f and with predicate symbols for all closed subsets of C^n as n varies. We call this language L_C. Notice that in this set up L_C has cardinality at least that of C.

So (M, f, C) is naturally a (first order) L_C-structure.

Logic comes into the picture in two ways. First by restricting the L_C-structures of interest to those where the interpretation of the second sort is precisely C. And secondly by restricting appropriately the class of L_C-formulas. We will call these (C, L)-structures (as already mentioned above) and (C, L)-formulas, which are to be defined. We may write a (C, L)-structure (M, f, C) as just (M, f).

Definition 2.2.12. A (C, L)-term is either an L-term, or something of the form $f(t)$ where t is an L-term. We call the former an L-valued (C, L)-term and latter a C-valued (C, L) term.

Definition 2.2.13. The class of (C, L)-formulas is the smallest set of L_C formulas which

(i) contains all L-formulas,
(ii) contains all formulas of the form $(s_1, ..., s_n) \in D$ where the s_i are C-valued (C, L)-terms, and D is a closed subset of C^n,
(iii) is closed under $\wedge$, $\vee$,
(iv) and is closed under $\forall x$ and $\exists x$ where x is a variable of the L-sort.

So notice that negation only appears inside L-formulas. Modulo this we are precisely in a positive logic setting. Also free variables in (C, L)-formulas are all of the L-sort. By a (C, L)-sentence we mean a (C, L)-formula without free variables.

A (C, L) structure is also an L_C structure, and a (C, L)-formula is an

L_C-formula, there is no need for a separate definition of truth for (C, L)-structures and formulas. It is just induced from truth of L_C-formulas in L_C-structures.

As in Henson's positive bounded logic we will need to deal with approximations and approximate truth. (This will not be needed in official continuous logic, as approximation will be built in to the choice of quantifiers.)

First, by an approximation to a closed set $D \subseteq C^n$ we mean a closed neighbourhood D' of D, namely a closed $D' \subseteq C^n$ such that $D \subseteq U \subseteq D'$ for some open U.

And by an approximation to a CL-formula φ we mean a CL-formula obtained from φ by replacing every occurrence of any closed set D in φ by some approximation to it. (One can also define this formally and inductively, as in Section 2.2.2.)

Definition 2.2.14.

(i) Let $\Gamma(\bar{x})$ be a collection of (C, L)-formulas which is closed under conjunctions. Then Γ^+ denotes the collection of all approximations of formulas in Γ.

(ii) We say that $\bar{a}$ approximately satisfies $\Gamma(\bar{x})$ in a (C, L)-structure (M, f), written $(M, f) \models_\sim \Gamma(\bar{a})$, if $(M, f, C) \models \varphi(\bar{a})$ for all $\varphi(\bar{x}) \in \Gamma^+$.

(iii) A set $\Gamma(\bar{x})$ of (C, L)-formulas is finitely approximately satisfiable if every finite subset of Γ^+ is satisfied in some (C, L)-structure.

As in Proposition 2.2.10 we have the compactness theorem.

Proposition 2.2.15. *Let Σ be a set of (C, L)-sentences which is finitely approximately satisfiable. Then Σ has a (C, L)-model.*

Proof. Like the proof of Proposition 2.2.10. Here is a sketch. First by the usual first order compactness theorem there is a (saturated) L_C-structure (M, f^*, C^*) which is a model of Σ^+. As C is a compact space and C^* is an elementary extension of C in the language with predicates for all closed subsets of the various C^n's, we have a canonical standard part map $st : C^* \to C$. Let $f = st \circ f^* : M \to C$. So (M, f) is a (C, L)-structure. Then first check that $(M, f) \models \Sigma^+$. Then using saturation of the L_C-structure (M, f^*, C^*) show that actually $(M, f) \models \Sigma$. □

There is an obvious notion of (M, f) being an approximate elementary substructure of (N, g).

A (C, L)-structure (M, f) is said to be κ-saturated if whenever $\Sigma(\bar{x})$ is s set of (C, L) formulas with parameters from a subset of M of cardinality $< \kappa$ and $\Sigma(\bar{x})$ is finitely approximately satisfiable in (M, f) then $\Sigma(\bar{x})$ is outright satisfiable (or realized) in (M, f).

As in Proposition 2.2.15, we obtain:

Proposition 2.2.16. *For any κ, any (C, L)-structure has a κ-saturated approximate elementary extension. Moreover for infinite κ, if (M, f) is κ-saturated, $\varphi(\bar{x})$ is a (C, L)-formula and $\bar{a}$ a tuple from M, then $(M, f) \models_{\sim} \varphi(\bar{a})$ if and only if $(M, f) \models \varphi(\bar{a})$.*

By a (C, L)-theory we mean a collection of (C, L)-sentences which is consistent, namely has a (C, L)-model. We call a (C, L)-theory complete if it is a maximal (C, L)-theory (i.e. maximal consistent). Then one can show that a (C, L)-theory is complete if and only if it is the approximate (C, L)-theory of some (C, L)-structure (M, f), namely the collection of (C, L)-sentences σ such that $(M, f) \models_{\sim} \sigma$.

Let us now discuss type spaces. We fix a complete (C, L)-theory T, and an ω-saturated model (M, f) of T (so truth and approximate truth coincide in (M, f)). Fix a tuple $\bar{x}$ of variables and a tuple $\bar{a}$ from M of the same length as $\bar{x}$. By the type of $\bar{a}$ in (M, f) ($tp_{(M,f)}(\bar{a})$) we mean the collection of (C, L)-formulas $\varphi(\bar{x})$ such that $(M, f) \models \varphi(\bar{a})$. The space of $\bar{x}$-types of T, $S_{\bar{x}}(T)$ is the collection of all such types, equipped with the topology whose basic closed sets are given by (C, L)-formulas $\varphi(\bar{x})$. Namely a basic closed set in $S_x(T)$ is the collection of types containing a given (C, L)-formula.

$S_{\bar{x}}(T)$ will be a compact Hausdorff space, but not necessarily a Stone space. This is typical of continuous logic. It is quite natural to ask, if (M, f) is a (C, L)-structure, what is the *induced first order structure* on M? Well let T be the (approximate) (C, L)-theory of (M, f), and add predicates to various Cartesian powers of M corresponding to clopen subsets of the appropriate type spaces $S_{\bar{x}}(T)$. This will be the induced first order structure. Let us connect this with Example 2.2.4 above (without proof).

Remark 2.2.17. Let L be a (first order) language and M an L-structure, and T the theory of M. Let C be a compact Hausdorff space and φ a C-valued formula in 1-variable, giving rise to a function $f : M \to C$. Consider the (C, L)-structure (M, f). Then there is no additional induced first order structure on M.

So given a (C, L)-structure (M, f) it is also natural to ask whether this is a *proper* expansion of M as opposed to f being already given by a C-valued formula of $Th(M)$. We give an example below.

Let us first define stability in a way consistent with Example 2.2.4.

Definition 2.2.18.

(i) Let now T be a complete (C,L)-theory, and $\varphi(x,y)$ a (C,L)-formula in tuples of variables x,y. $\varphi(x,y)$ is *stable* (for T), if working in a saturated model (M,f) of T, if $(a_i,b_i)_{i<\omega}$ is an indiscernible sequence, and $(M,f)\models\varphi(a_i,b_j)$ for $i<j$ then also $(M,f)\models\varphi(a_i,b_j)$ for $i>j$.

(ii) The complete (C,L)-theory is stable if every (C,L)-formula $\varphi(x,y)$ is stable.

We complete this section with a description (without proof) of results from [CP23]. Among the important examples of stable first order structures are what are known as *abelian structures*, namely an abelian group $(A,+)$ together with a collection of predicates for various subgroups of various Cartesian powers of A (including of course equality). If M is such a first order structure then not only is $Th(M)$ stable but also is what is known as 1-based, and conversely any 1-based stable groups is essentially. (See Definition 1.3.62 of the chapter on Stability Theory.)

Proposition 2.2.19. *Let $M=(A,+,...)$ be an abelian structure in language L. Let C be a compact Hausdorff group, and f a homomorphism from $(A,+)$ to C. Then the approximate (C,L)-theory of (M,f) is stable (and even 1-based after making the suitable definition).*

Example 2.2.20. Let L be the language of abelian groups. Let f be a homomorphism from $(\mathbb{Z},+)$ to the circle group S^1 obtained by mapping 1 to an irrational rotation. Then the (approximate) (S^1,L)-theory of $((\mathbb{Z},+),f)$ is stable.

Moreover this does not "come from" a stable first order structure. Namely there is no first order language L' extending L, and expansion of $(\mathbb{Z},+)$ to a stable L' structure M say, such that f is (comes from) a S^1-valued formula in the sense of $Th(M)$.

The reason is roughly as follows: Suppose f were an S^1-valued formula in the sense of $Th(M)$. Take a saturated elementary extension (M',f') of the (S^1,L') structure (M,f). Then f' is still a S^1-valued formula of $Th(M')$ and is moreover a homomorphism onto S^1. But then $ker(f')$ will be a type-definable in L' subgroup of M' of bounded index, which by stability of $Th(M')$ has to be an intersection of finite index L'-definable subgroups of M'. Some further arguments imply that S^1 should be a profinite group (as a compact group) which gives a contradiction.

2.3 "Official" Continuous Logic

We will be going through [BBHU08], [BU10]. We have discussed a few formalisms for continuous logic, and now we will fix one which is used by many researchers who work with continuous logic.

2.3.1 *Structures and vocabulary*

Recall: a **metric space** is a set M equipped with a map $d : M \times M \to \mathbb{R}_{\geqslant 0}$ satisfying

- $d(x, y) = d(y, x)$
- $d(x, y) = 0 \Leftrightarrow x = y$
- $d(x, y) + d(y, z) \geqslant d(x, z)$.

A metric space is a topological space where the basic open neighbourhoods of $a \in M$ are $B_\epsilon(a) = \{x : d(x, a) < \epsilon\}$.

If $(M_1, d_1), \ldots, (M_n, d_n)$ are metric spaces, then so is their product $(M_1 \times \cdots \times M_n, d)$ where $d(\bar{x}, \bar{y}) = \max\{d_1(x_1, y_1), \ldots, d_n(x_n, y_n)\}$, and this product has the product topology of the topologies of the original metric spaces. Note that boundedness and completeness are also preserved by metric space products.

$F : M_1 \to M_2$ is uniformly continuous if for all ϵ there is δ such that $d_1(x, y) < \delta \Rightarrow d_2(F(x), F(y)) < \epsilon$ for all $x, y \in M_1$. If we can choose $\delta = \Delta(\epsilon)$ to be a function of ϵ, we call Δ a modulus of uniform continuity. See appendix to Section 2 of [BBHU08] for remarks on such moduli.

M is **complete** if every Cauchy sequence converges.

M is **bounded** if there is $N \in \mathbb{N}$ such that $d(x, y) \leqslant N$ for all $x, y \in M$.

Definition 2.3.1 (Metric Space Structure). **A metric space structure** is a complete, bounded metric space (M, d), bounded by $N = 1$, equipped with

- a family of predicates $\{R_i : i \in I\}$, where by a predicate R we mean a uniformly continuous function from some M^n to a bounded interval in $\mathbb{R}$ which we take to be $[0, 1]$;
- a family of functions $\{F_j : j \in J\}$, where by a function we mean $F : M^n \to M$ which is uniformly continuous (constants are treated as 0-ary functions).

This is a 1-sorted metric space structure. There is a natural definition of a many-sorted metric space structure. As with FOL, if $\mathcal{M} = (M, d,$

$(R_i)_I, (F_j)_J)$ is a one-sorted structure, we will often refer to $\mathcal{M}$ as its underlying set M.

Example 2.3.2.

- An ordinary FOL structure is a special case, where $d(x, y) = 1$ if $x \neq y$. The predicates have value 0 (true) and 1 (false).
- Henson's normed space structures $(V, +, ||\cdot||, \dots)$, which we modify by considering only $V_1 = \{x \in V : ||x|| \leqslant 1\}$, restricting all functions $V \to \mathbb{R}$ to V_1 as new *predicates*, and restricting $+$ to a suitable domain. The metric here is $d(x, y) = ||x - y||$. Note that the real sort is not included here because it is part of the logic. On the other hand we could consider $(V, +, ||\cdot||, \dots)$ as a many-sorted metric structure by having sorts for the balls of diameters n, together with the natural inclusion maps between sorts.
- Probability spaces. Consider $(\Omega, \mathcal{B}, \mu)$, where Ω is a set, $\mathcal{B}$ is a σ-algebra on Ω (i.e. a family of subsets of Ω containing Ω, $\varnothing$, closed under countable unions, countable intersections, and complementation), and μ is a countably additive probability measure on $\mathcal{B}$ (i.e. $\mu(\Omega) = 1$, $\mu(\varnothing) = 0$, and if (A_i) is a countable family of disjoint subsets with $A = \bigcup_i A_i$, then $\mu(A) = \Sigma_i \mu(A_i)$). Let E be the equivalence relation on $\mathcal{B}$ given by $E(A_1, A_2)$ iff $\mu(A_1 \Delta A_2) = 0$, where Δ is the symmetric difference. Let $\hat{\mathcal{B}} = \mathcal{B}/E$. The metric structure is then $(\hat{\mathcal{B}}, 0, 1, c, \bigcup, \bigcap, \hat{\mu})$, where $d(A/E, B/E) = \mu(A \Delta B)$, $0 = \varnothing/E$, $1 = \mathcal{B}/E$, c means complementation, and $\hat{\mu}$ is the induced measure $\hat{\mathcal{B}} \to [0, 1]$. (By countable additivity, this metric is guaranteed to be complete.)

We now develop the syntax for metric space structures. We have an obvious vocabulary/language L for a metric space structure $\mathcal{M} = (M, d, (R_i)_i, (F_j)_j)$. For each predicate $R : M^n \to [0, 1]$, we have a *predicate symbol* P equipped with an arity $a(P) = n$, and a modulus of uniform continuity Δ_P. Likewise for each function $F : M^n \to M$ we have a function symbol f, $a(f) = n$, and again have an associated modulus Δ_f. (Think of the moduli like arity: *fixed* functions which implicitly part of the syntax, but not actually symbols in the language.) So M is an L-structure, and via the moduli, the uniform continuity conditions are *built into the language.*

Given a vocabulary L, we can consider **L-prestructures**. By a pseudometric d on a set M we mean that d satisfies all the metric space axioms except $d(x, y) = 0 \Rightarrow x = y$. So instead of forcing equality, d induces an equivalence relation.

By an L-prestructure, we mean a pseudometric space of diameter $\leqslant 1$ equipped with interpretations of predicate and function symbols of L satisfying the arity and uniform continuity data with respect to the pseudometric.

Given a pseudometric space (M_0, d_0), we can obtain an actual metric space (M, d) by quotienting out by the equivalence relation $d_0(x, y) = 0$. By the uniform continuity conditions, the distinguished predicates and functions of M_0 are well defined on M after this quotient, with the same moduli. Consider the quotient map $\pi : M_0 \to M$. We can see that $F(x) = F(y)$ if $\pi(x) = \pi(y)$. Now pass to the completion $(\bar{M}, \bar{d}, \dots)$ of (M, d) to get an L-structure.

2.3.2 *Syntax and semantics*

In the previous section, we defined a language $\mathcal{L}$ in first order continuous logic which consisted of function symbols and predicate symbols. Each predicate symbol P is equipped with an arity $a(P)$ and a modulus of uniform continuity Δ_P. Given an $\mathcal{L}$ (pre)structure $\mathcal{M}$, the interpretation of P in M, $P(\mathcal{M})$, is a function from $M^{a(P)} \to [0, 1]$ which is uniformly continuous with respect to the modulus Δ_P. Similarly, each function symbol f is equipped with an arity $a(f)$ and a modulus of uniform continuity Δ_f. The interpretation of f in $\mathcal{M}$, $f(\mathcal{M})$, is a function from $M^{a(f)}$ to M which is uniformly continuous with respect to Δ_f.

There is a natural notion of isomorphism for $\mathcal{L}$ (pre)structures: given two $\mathcal{L}$ (pre)structures $\mathcal{M}_0$ and $\mathcal{M}_1$, an isomorphism between them is a bijection θ from M_0 to M_1 such that for all predicate symbols P and all $\bar{a} \in M_0^{a(P)}$, $P(\mathcal{M}_0)(\bar{a}) = P(\mathcal{M}_1)(\theta(\bar{a}))$. Similarly for all function symbols f and all $\bar{a} \in M_0^{a(f)}$, $f(\mathcal{M}_0)(\bar{a}) = f(\mathcal{M}_1)(\theta(\bar{a}))$. Note that as the (pseudo)metric d on our (pre)structures is just a special predicate symbol, this implies that a bijection is an isometry.

Remark. The difference between Henson's formalism and "official" continuous logic is superficial. Henson has another sort for $\mathbb{R}$, and his functions to $\mathbb{R}$ are instead treated as $\mathbb{R}$-valued predicates.

Note that the cardinality of $\mathcal{L}$ is the sum of the cardinalities of the predicate symbols and function symbols.

2.3.2.1 *Completions of metric space*

Given the fact that $\mathcal{L}$ structures are complete but $\mathcal{L}$ prestructures need not be, we briefly comment on the completions of metric spaces. Given a metric space (M, d), we define an equivalence relation $\sim$ on the set of Cauchy sequences of elements of M. (Recall a sequence $(a_n)_{n\in\omega}$ is Cauchy if for every $\epsilon > 0$ there exists $N \in \omega$ such that for all $n, m > N$, $d(a_n, a_m) < \epsilon$.) Specifically, $(a_n)_{n\in\omega} \sim (b_n)_{n\in\omega}$ if and only if for all $\epsilon > 0$ there exists an $N \in \omega$ such that for all $n, m > N$, we have $d(a_n, b_m) < \epsilon$.

Let $\hat{M} = M/\sim$ and let $\hat{d}(\hat{a}, \hat{b}) = \lim_n d(a_n, b_n)$ for $\hat{a}$ the equivalence class of $(a_n)_{n\in\omega}$ and $\hat{b}$ the equivalence class of $(b_n)_{n\in\omega}$. Note that this limit exists, as for any $\epsilon > 0$ there exists N such that $d(a_n, a_m) < \epsilon$, $d(b_n, b_m) < \epsilon$, and $d(a_n, b_n) > \hat{d}(\hat{a}, \hat{b}) - \epsilon$ for all $n, m > N$. Therefore $d(a_m, b_m) \leqslant d(a_n, a_m) + d(a_n, b_n) + d(b_n, b_m) \leqslant 3\epsilon$, so $\hat{d}(\hat{a}, \hat{b}) = \limsup_n d(a_n, b_n)$.

Additionally, this is well defined because if $(x_n)_{n\in\omega}$ and $(y_n)_{n\in\omega}$ are different representatives from $\hat{a}$ and $\hat{b}$ respectively, for any $\epsilon > 0$ there is some N such that for all $n > N$ we have $d(x_n, a_n) < \epsilon$ and $d(b_n, y_n) < \epsilon$ by the definition of $\sim$, thus $d(x_n, y_n) \leqslant d(x_n, a_n) + d(a_n, b_n) + d(b_n, y_n) < d(a_n, b_m) + 2\epsilon$ by the triangle inequality and therefore $\lim_n d(a_n, b_n) = \lim_n d(x_n, y_n)$.

Now we shall see that $\hat{M}$ is complete. Let $(\hat{a}_n)_{n\in\omega}$ be a Cauchy sequence in $\hat{M}$ with $\hat{a}_n$ the equivalence class of the Cauchy sequence $(a_{n,m})_{m\in\omega}$. Consider $(a_{n,n})_{n\in\omega}$. First, note that it is Cauchy: If $\epsilon > 0$, there exists N such that $\hat{d}(\hat{a}_n, \hat{a}_m) < \epsilon$ for all $n, m > N$. Therefore $\lim_x d(a_{n,x}, a_{m,x}) < \epsilon$ by the above observation. Thus there is some $M \geqslant N$ such that $d(a_{n,x}, a_{m,x}) < \epsilon$ for all $x > M$, so taking n, m, x greater than the modulus for ϵ for the sequences $(a_{n,x})_{x\in\omega}$ and $(a_{m,x})_{x\in\omega}$ respectively, we get $d(a_{n,n}, a_{m,m}) \leqslant d(a_{n,n}, a_{n,x}) + d(a_{n,x}, a_{m,x}) + d(a_{m,x}, a_{m,m}) < 3\epsilon$. Thus $(a_{n,n})_{n\in\omega}$ is Cauchy.

Furthermore, $(\hat{a}_n)_{n\in\omega}$ converges to the equivalence class $\hat{b}$ of $(a_{n,n})_{n\in\omega}$: As $(\hat{a}_n)_{n\in\omega}$ is Cauchy, for all $\epsilon > 0$ there exists some N such that for all $n, m > N$, $\hat{d}(\hat{a}_n, \hat{a}_m) = \lim_k d(a_{n,k}, a_{m,k}) < \epsilon$. Thus there is $M \geqslant N$ with $d(a_{n,k}, a_{m,k}) < \epsilon$ for all $k > M$, so we have $d(a_{n,k}, a_{k,k}) < \epsilon$. Therefore $\epsilon \geqslant \limsup_k d(a_{n,k}, a_{k,k}) = \hat{d}(\hat{a}_n, \hat{b})$, so $(\hat{a}_n)_{n\in\omega}$ converges to $\hat{b}$ as desired.

Finally, note that we can embed (M, d) into $(\hat{M}, \hat{d})$ via the map which takes a to $(a)_{n\in\omega}$. Then for each predicate symbol P define $P(\hat{a}_1, \ldots, \hat{a}_{a(P)}) = \lim_n P(a_{1,n}, \ldots, a_{a(P),n})$ and for each function symbol f define $f(\hat{a}_1, \ldots, \hat{a}_{a(f)})$ to be the equivalence class of $(f(a_{1,n}, \ldots, a_{a(f),n}))_{n\in\omega}$. Uniform continuity for $P(\hat{M})$ and $f(\hat{M})$ follows immediately from uniform continuity for $P(M)$ and $f(M)$ respectively and the fact that for any two equivalence classes $\hat{a}$ and $\hat{b}$ with $\hat{d}(\hat{a}, \hat{b}) < \delta$, we can choose representatives $(a_n)_{n\in\omega}$ and $(b_n)_{n\in\omega}$ with $d(a_n, b_n) < \delta$ for all n.

2.3.2.2 *Syntax*

In standard first order logic, we have to distinguish between the nonlogical symbols, such as predicate symbols, function symbols, and constants and the logical symbols, such as parentheses, commas, and, or, and quantifiers. We do the same in continuous logic:

- The nonlogical symbols are the predicate symbols and function symbols from $\mathcal{L}$.
- The logical symbols are parentheses, commas, variable symbols, propositional connectives, and quantifiers.

Variable symbols are $v_0, v_1, \ldots$ indexed by the natural numbers. If desired, uncountably many variable symbols can be used. There is a propositional connective symbol for each continuous function from $[0,1]^n \to [0,1]$ for all n. The quantifier symbols are inf and sup, which will be interpreted as the infimum and supremum respectively.

d is considered a nonlogical symbol as it is just a binary predicate symbol, albeit a special one present in every language $\mathcal{L}$. Conversely, continuous logic does not have the equality symbol. However, it is possible to choose a modulus of continuity in the language to force the metric to be discrete, i.e. force it to be equality.

Definition 2.3.3. We define the syntax of our language starting with terms. These are defined recursively.

- All variable symbols are terms.
- If $t_1, \ldots, t_{a(f)}$ are terms for some function symbol f, then $f(t_1, \ldots, t_{a(f)})$ is also a term.
- All terms are built up using these two rules.

Using terms, we next define formulas recursively.

- If P is a predicate symbol and $t_1, \ldots, t_{a(P)}$ are terms, then $P(t_1, \ldots, t_{a(P)})$ is an atomic formula.
- If $u : [0,1]^n \to [0,1]$ is a propositional connective and $\varphi_1, \ldots, \varphi_n$ are formulas, then $u(\varphi_1, \ldots, \varphi_n)$ is a formula.

- If x is a variable symbol and φ is a formula, then $\inf_x \varphi$ and $\sup_x \varphi$ are formulas.
- All formulas are built up using these rules.

We next use recursion to define free variables.

- Every variable appearing in an atomic formula is free.
- If $u : [0,1]^n \to [0,1]$ is a propositional connective and $\varphi_1, \ldots, \varphi_n$ are formulas, then the free variables in $u(\varphi_1, \ldots, \varphi_n)$ are all of the variables which appear free in at least one of $\varphi_1, \ldots, \varphi_n$.
- If x is a variable symbol and φ is a formula, then the free variables in $\inf_x \varphi$ and $\sup_x \varphi$ are the free variables in φ minus x.

A formula is said to be quantifier free if it is built up without using $\inf_x$ or $\sup_x$ for any variable symbol x. Every variable appearing in a quantifier free formula is free.

An $\mathcal{L}$ sentence is an $\mathcal{L}$ formula with no free variables.

2.3.2.3 *Semantics*

Definition 2.3.4. Let $\mathcal{M}$ be an $\mathcal{L}$ prestructure. By $\varphi(x_1, \ldots, x_n)$, we mean φ is a formula whose free variables are included in $x_1, \ldots, x_n$. However, we do not require that all of $x_1, \ldots, x_n$ appear free. Given such a formula $\varphi(x_1, .., x_n)$, a prestructure M, and an assignment of elements a_i in M to variable x_i, $\varphi(\mathcal{M})(\bar{a})$ will be in $[0,1]$. Defined inductively below.

We first define the interpretation of terms in $\mathcal{M}$ inductively.

- For variable symbols x, the interpretation of x in $\mathcal{M}$ under the assignment of a to x, denoted by $x(\mathcal{M})(a)$ is the element a.
- Given a function symbol f and terms $t_1, \ldots, t_{a(f)}$ whose variables are included in $x_1, \ldots, x_n$, the interpretation of $f(t_1, \ldots, t_{a(f)})$ in $\mathcal{M}$ under the assignment $\bar{a}$, denoted by $f(t_1, \ldots, t_{a(f)})(\mathcal{M})(\bar{a})$ is defined to be $f(\mathcal{M})(t_1(\mathcal{M})(\bar{a}), \ldots, t_{a(f)}(\mathcal{M})(\bar{a}))$.

We now inductively define the interpretation $\varphi(x_1, \ldots, x_n)(\mathcal{M})(\bar{a})$, of a formula $\varphi(x_1, ..., x_n)$ in $\mathcal{M}$ under interpretation $\bar{a}$:

- For atomic formulas $\varphi = P(t_1, \ldots, t_{a(P)})$ with the collective variable symbols included in $x_1, \ldots, x_n$, we define

$$\varphi(\mathcal{M})(\bar{a}) = P(\mathcal{M})(t_1(\mathcal{M})(\bar{a}), \ldots, t_{a(P)}(\mathcal{M})(\bar{a})).$$

- If $\varphi_1, \ldots, \varphi_m$ are formulas whose collective variable symbols are included in $x_1, \ldots, x_n$ and $u : [0,1]^m \to [0,1]$ is a propositional connective, then

$$u(\varphi_1, \ldots, \varphi_m)(\mathcal{M})(\bar{a}) = u(\varphi_1(\mathcal{M})(\bar{a}), \ldots, \varphi_m(\mathcal{M})(\bar{a}).$$

- If φ is a formula whose variable symbols are included in $x_1, \ldots, x_n, y$, then

$$\sup_y \varphi(\mathcal{M})(\bar{a}) = \sup_b \{\varphi(\mathcal{M})(\bar{a}, b) : b \in M\}.$$

Similarly,

$$\inf_y \varphi(\mathcal{M})(\bar{a}) = \inf_b \{\varphi(\mathcal{M})(\bar{a}, b) : b \in M\}.$$

Definition 2.3.5.

1. Let $\mathcal{M}, \mathcal{N}$ be $\mathcal{L}$ prestructures. We say $\mathcal{M}$ and $\mathcal{N}$ are elementarily equivalent if $\sigma(\mathcal{M}) = \sigma(\mathcal{N})$ for all $\mathcal{L}$ sentences σ.
2. A morphism from an $\mathcal{L}$ prestructure $\mathcal{M}$ to an $\mathcal{L}$ prestructure $\mathcal{N}$ is a map $h : M \to N$ such that for all $h(f(\mathcal{M})(\bar{a})) = f(\mathcal{N})(h(\bar{a}))$ for all function symbols f and $\bar{a} \in M^{a(f)}$, and $P(\mathcal{M})(\bar{a}) = P(\mathcal{N})(h(\bar{a}))$ for all predicate symbols P and $\bar{a} \in M^{a(P)}$. Note that this applies in particular for the special predicate symbol d, so if $\mathcal{M}$ is an $\mathcal{L}$ structure then h must be an embedding.
3. Such a morphism is called elementary if it preserves all $\mathcal{L}$ formulas in similar fashion, i.e. for any $\mathcal{L}$ formula $\varphi(x_1, \ldots, x_n)$, $\varphi(\mathcal{M})(\bar{a}) = \varphi(\mathcal{N})(h(\bar{a}))$ for all $\bar{a} \in M^n$.

Remark 2.3.6. Terms and formulas of $\mathcal{L}$ also have moduli of uniform continuity in $\mathcal{L}$ prestructures, just depending on $\mathcal{L}$. Every $\mathcal{L}$ prestructure in the same fashion as $\mathcal{L}$ function symbols and predicate symbols. For more information, see the Appendix to Section 2 in [BBHU08].

We shall now discuss propositional connectives in more detail. Recall that a propositional connective u is any continuous function from $[0,1]^n$ to $[0,1]$ for any $n \in \omega$. However, this leads to continuum-many propositional connectives, which is not ideal because we often want our language to be countable. We shall see that we can in fact restrict down to finitely many propositional connectives. In the case of FOL the analogue is considering all functions from $\{0,1\}^n$ to $\{0,1\}$ (n varying) and noting that we can get by with the usual symbols, $\wedge$, $\vee$, $\neg$.

First let us give some examples of propositional connectives in continuous logic:

- Any constant function defined for $r \in [0, 1]$.
- $\neg x$, representing the continuous function $1 - x$.
- $x \wedge y$, "x and y", representing the two-variable continuous function $max(x, y)$.
- $x \vee y$, "x or y", representing the two-variable continuous function $min(x, y)$.
- $x \dot{-} y$, defined to be $x - y$ if $x \geqslant y$ and 0 otherwise.

Let $\mathcal{F} = \{F_n : n \in \omega\}$ be a system of collections F_n of continuous functions from $[0, 1]^n$ to $[0, 1]$. Then by $\overline{\mathcal{F}}$ we denote $\{F'_n : n \in \omega\}$, where the F'_n's are obtained from the F_n's by closing under projections and compositions. We say $\mathcal{F}$ is full if each F'_n is dense in the space of continuous functions from $[0, 1]^n$ to $[0, 1]$ where the topology is the compact-open topology, or equivalently the uniform convergence topology. The topology defined by the supremum metric is also equivalent to the compact open topology, i.e. the metric defined via $d(f, g) = \sup\{|f(x) - g(x)| : x \in [0, 1]^n\}$.

The compact open topology is defined using closed sets $K \subseteq [0, 1]^n$ and open sets $U \subseteq [0, 1]$. Then $C(K, U)$ is the set of continuous functions from $[0, 1]^n$ to $[0, 1]$ with $f(K) \subseteq U$. Then the finite intersections of $C(K, U)$'s form a basis for the compact open topology.

A sequence of functions $(f_m)_{m \in \omega}$ from $[0, 1]^n$ to $[0, 1]$ uniformly converges to f if for all $\epsilon > 0$ there exists an N such that for all $n > N$ and $x \in [0, 1]^n$, $|f_n(x) - f(x)| < \epsilon$. Then the closed sets in the uniform convergence topology are exactly those that are closed under uniform convergence, i.e. those sets such that all uniformly convergent sequences have their limit functions also in the set.

Example 2.3.7. The systems $\mathcal{F}$ defined by the following F_n are full.

- F_0 contains all rational constant functions, $F_2 = \{x \dot{-} y\}$, and all other F_n are empty.
- $F_1 = \{\neg x, \frac{x}{2}\}$, $F_2 = \{\dot{-}\}$, and all other F_n are empty.
- $F_0 = \{0, 1\}$, $F_1 = \{\frac{x}{2}\}$, $F_2 = \{\dot{-}\}$, and all other F_n are empty.

For details regarding why these are full, see [BBHU08] and [BU10].

Lemma 2.3.8. *(See Theorem 6.3 [BBHU08]) Assume $\mathcal{F}$ is a full system of connectives. By an $\mathcal{F}$-restricted formula we mean an $\mathcal{L}$-formula constructed as usual by induction, but where in the clause involving $u(\varphi_1, ..., \varphi_n)$, only* *connectives $u \in F_n$ are allowed. For any $\mathcal{L}$-formula $\varphi(\bar{x})$ and $\varepsilon > 0$ there is an $\mathcal{F}$ restricted $\mathcal{L}$ formula $\psi(\bar{x})$ such that for all $\mathcal{L}$ structures M and $\bar{a} \in M^n$, $|\varphi(M)(\bar{a}) - \psi(M)(\bar{a})| < \varepsilon$.*

As was discussed in the section on introductory material and historical background, in the continuous setting truth and approximate truth are closely linked. The lemma guarantees that a full system of connectives yields "enough" $\mathcal{F}$-restricted formulae to approximate all $\mathcal{L}$-formulae. It is hard to obtain good bounds on the number of $\mathcal{L}$ formulae due to the number of non-logical symbols. However, (for full $\mathcal{F}$) it is manageable to obtain bounds on the number of $\mathcal{F}$-restricted formulae.

In 6.9 of [BBHU08] it is stated that for $\mathcal{F}$ as defined in bullet 3 of Example 2.3.7 above, any $\mathcal{F}$-restricted formula is equivalent to an $\mathcal{F}$-restricted formula in prenex form.

A remark about the quantifiers *inf* and *sup*. Notice that FOL is the special case of continuous logic where the metric is discrete and all formulas are $\{0, 1\}$-valued where 0 stands for true and 1 for false. So let M be a FOL structure in this sense and $\varphi(\bar{x}, y)$ a formula of the relevant language. What does it mean that $inf_y \varphi(\bar{x}, y)$ has value 0 at $\bar{a}$? Well it just means that $M \models \exists y \varphi(\bar{a}, y)$. Likewise $sup_y \varphi(\bar{x}, y)$ has value 0 at $\bar{a}$ means $M \models \forall y \varphi(\bar{a}, y)$.

Remark 2.3.9. Let X be a compact Hausdorff space and let X^* be the set of closed subsets of X. The Vietoris Topology on X^* has the following basis of open sets: given $U, V_1, ..., V_n$ open in X, we take $\{Y \in X^* : Y \subset U, Y \cap V_i \neq \varnothing, i = 1, ..., n\}$ as a basic open set. This makes X^* compact and Hausdorff as well.

Back in continuous logic, consider an L-formula $\varphi(\bar{x}, y)$ and $\mathcal{L}$-structure (or prestructure) M and $\bar{a} \in M^n$ consider the set $\{\varphi(M)(\bar{a}, b) : b \in M\} \subseteq [0, 1]$. Let $Z = \overline{\{\varphi(M)(\bar{a}, b) : b \in M\}} \subseteq [0, 1]$ denote the closure of this set. Then $\inf\{\varphi(M)(\bar{a}, b) : b \in M\} = \min(Z)$. Note that $\min : [0, 1]^* \to [0, 1]$ is continuous. An analogous statement holds for max and sup.

So the quantifiers inf and sup arise from continuous functions from $[0, 1]^*$ to $[0, 1]$. Indeed, other quantifiers q could be introduced by considering other such continuous functions. So for example $q_y(\varphi(\bar{x}, y)(M)(\bar{a})) = q\overline{\{\varphi(M)(\bar{a}, b) : b \in M\}}$. For more on this perspective see [CK85]. The quantifiers inf and sup are supposed to be enough in some suitable sense to be discussed later in the section on definability.

Proposition 2.3.10. *(Proposition 2.10 [BBHU08]) Let M be an $\mathcal{L}$-prestructure. Consider the $\mathcal{L}$-structure $\hat{M}$ obtained from M by first* *quotienting by a suitable equivalence relation E, and then taking the completion $M \to M/E \to \hat{M}$. The map $h : M \to \hat{M}$ is a morphism of prestructures, and in fact h is elementary, i.e. it preserves all formulas.*

Proof. Induction on formulas. □

Possibly some of the expressions used in [BBHU08] are not so optimal or memorable. Here is an example:

Definition 2.3.11. By an $\mathcal{L}$-condition we mean an expression of the form $\varphi = 0$. Where φ is an $\mathcal{L}$-formula. When φ is a sentence, the expression is called a closed or sentential condition.

If M is an $\mathcal{L}$-structure, $\bar{a} \in M^n$, and $\varphi(\bar{x})$ an $\mathcal{L}$-formula then the condition "$\varphi = 0$" is either true or false for $\bar{a}$.

Remark 2.3.12. As any constant function $r \in [0,1]$ is a connective and $|x - y|$ is a connective, then for any $\mathcal{L}$-formula $\varphi(\bar{x})$, $|\varphi(\bar{x}) - r|$ is a formula. The condition $|\varphi(\bar{x}) - r| = 0$ says $\varphi = r$. Likewise using $\dot{-}$ we see $\varphi \leqslant r$ and $\varphi \geqslant r$ and $\varphi \leqslant \psi$ and $\varphi \geqslant \psi$ are all conditions.

An $\mathcal{L}$-theory T is defined to be a set of closed conditions, and a model M of T is an $\mathcal{L}$-structure M such that $\sigma(M) = 0$ for each condition $\sigma = 0$ in T. Likewise we have $Th(M)$ for any $\mathcal{L}$-structure, a *complete* theory. Note that $Th(M)$ consists of giving the value in M of any $\mathcal{L}$-sentence σ.

In the spirit of redoing elementary model theory in the continuous setting, we have:

Lemma 2.3.13. *Let S be a dense subset of all $\mathcal{L}$-formulas in the sense that for every $\mathcal{L}$ formula $\varphi(\bar{x})$ and $\varepsilon > 0$ there is a formula $\psi(\bar{x}) \in S$ such that for all $\mathcal{L}$-structures M and $\forall \bar{a} \in M^n$, $|\varphi(M)(\bar{a}) - \psi(M)(\bar{a})| < \varepsilon$. Let M be an $\mathcal{L}$-substructure of N. Then the following are equivalent:*

1. *$M \prec N$*
2. *For each $\psi(\bar{x}, y) \in S$ and $\bar{a} \in M^n$*

$$\inf\{\psi(N)(\bar{a}, b) : b \in M\} = \inf\{\psi(N)(\bar{a}, b) : b \in N\}.$$

Proof. The proof is analogous to the FOL proof of the Tarski-Vaught Test, induction over the structure of the formulas. See Proposition 4.5 in [BBHU08]. □

2.3.3 *Ultraproducts and compactness*

This section assumes familiarity with filters and ultrafilters on a set I. The machinery of metric ultraproducts will be developed and used to prove compactness. Before the development of continuous logic, metric ultraproducts were a tool of functional analysts; appearing in the work of McDuff, Connes, Lindenstrauss and others.

Fact 2.3.14. *Let X be a compact Hausdorff space, I an indexing set, and $\mathcal{U}$ be an ultrafilter on I. Let $(a_i)_{i\in I}$ be an I-indexed sequence of points in X. Then there is a unique point $a \in X$ such that for every open neighborhood O of a, $\{i \in I : a_i \in O\} \in \mathcal{U}$. We call $a = \lim_{i\to\mathcal{U}} a_i$.*

Proof. The uniqueness of a follows from the Hausdorff property. Now we prove existence:

Consider X as a (FOL) structure with all subsets named as predicates. Given $(a_i)_{i\in I}$ and $\mathcal{U}$ as above, consider the quantifier free type $p(x)$ where $Y \in p(x)$ iff $\{i \in I : a_i \in Y\} \in \mathcal{U}$. Let $X \prec X^*$, and $b \in X^*$ realize $p(x)$. Note we assume p is not realized in X. By the compactness of X we have a surjective standard part map $\mathrm{st} : X^* \to X$. Let $a = \mathrm{st}(b)$ then any open neighborhood of a is named in $p(x)$. □

Remarks on Fact 2.3.14: The Hausdorffness of X gives that the standard part map is well defined on its domain and the compactness of X ensures that the domain is all of X^*. Surjectivity is immediate. Finally, Fact 2.3.14 characterizes compact Hausdorff spaces.

Let I be an index set and $\mathcal{U}$ be an ultrafilter on I. Fix a set of $\mathcal{L}$-structures (M_i, d_i) indexed by I where d_i is the metric on M_i. We can assume that the M_i all have bound 1. Define the metric ultraproduct $\prod_{\mathcal{U}} M_i$ as follows.

First let N_0 be the following $\mathcal{L}$-prestructure. The universe of N_0 is $\prod_{i\in I} M_i$. Define a pseudometric d on N_0 by $d(a,b) = \lim_{i\to\mathcal{U}} d_i(a_i, b_i)$. If F is an n-ary function symbol of $\mathcal{L}$ define $F(N_0)(a^1, ..., a^n) = ((F(M_i)(a_i^1, ..., a_i^n))_{i\in I} \in N_0$. If P is an n-ary predicate symbol of $\mathcal{L}$, define $P(N_0)(a^1, ..., a^n) = \lim_{i\to\mathcal{U}}(P(M_i)(a_i^1, ..., a_i^n)) \in [0,1]$.

N_0 is thus an $\mathcal{L}$-prestructure which will have bound 1. If the $F(M_i)$ have the appropriate moduli of uniform continuity then $F(N_0)$ will as well.

Let $N = \prod_{\mathcal{U}} M_i$ be the corresponding $\mathcal{L}$-structure N_0/E, where $E(x,y)$ holds if $d(x,y) = 0$. The completeness of N follows from the completeness of the M_i; we prove this claim below. We call N the ultraproduct of the M_i across $\mathcal{U}$.

Proof. We now prove the claim that (N, d) is a complete metric space; see [BBHU08] for another account.

Let $(x^k)_{k\geqslant 1}$ be a Cauchy sequence in N; in order to show N is complete we must find a limit for this sequence in N. To begin, we may "thin" the sequence such that the following condition is satisfied for each k: $d(x^k, x^{k+1}) < \frac{1}{2^k}$. We may choose a representative sequence $(x_i^k) \in N_0$ for each of our $x^k \in N$. For each $m \geqslant 1$, define the set $A_m = \{i \in I\colon d_i(x_i^k, x_i^{k+1}) < \frac{1}{2^k}, \forall k = 1, \cdots, m\}$.

We claim each $A_m \in \mathcal{U}$. To see this, fix k. By definition of the pseudometric on N_0, we have that $d(x^k, x^{k+1}) = \lim_{i\to\mathcal{U}} d_i(x_i^k, x_i^{k+1})$. Let $c = d(x^k, x^{k+1})$. Let O be an open neighborhood of c in $[0, 1]$ which is contained in $[0, \frac{1}{2^k}]$. Now, by definition of the limit along $\mathcal{U}$, we know that $\{i\colon d_i(x_i^k, x_i^{k+1}) \in O\} \in \mathcal{U}$. So in particular, $\{i\colon d_i(x_i^k, x_i^{k+1}) < \frac{1}{2^k}\} \in \mathcal{U}$. Thus the finite intersection of these sets for $k = 1, \cdots, m$ is in $\mathcal{U}$. But this set is precisely A_m, hence $A_m \in \mathcal{U}$.

It is clear that $A_1 \supseteq A_2 \supseteq \cdots$. We are now ready to construct $(y_i)_{i\in I} \in N_0$ whose equivalence class is the limit of our given sequence (x^k). If $i \notin A_1$, define $y_i \in M_i$ arbitrarily (recall this happens only for a "small" collection of i, i.e. a set not in the ultrafilter as $A_1 \in \mathcal{U}$). If $i \in A_m \backslash A_{m+1}$, then let $y_i = x_i^{m+1}$. Finally, if $i \in \bigcap_{m<\omega} A_m$, then $(x_i^m)_m$ is already a Cauchy sequence in M_i; in this case let $y_i \in M_i$ be the limit of that Cauchy sequence, which exists since M_i is complete.

We now check that the given sequence (y_i) is in fact the limit of the sequence. It is clear from the choice of y_i that for each m, $\{i\colon d_i(x_i^m, y_i) \leqslant \frac{1}{2^{m-1}}\} \in \mathcal{U}$. From this it follows that $\lim_{i\in\mathcal{U}} d_i(x_i^m, y_i) \leqslant \frac{1}{2^{m-1}}$. But then it follows from the definition of the distance d on N_0 that (y_i) is the limit of the sequence (x^k) in prestructure N_0, and by passing to the equivalence classes we get that it is the limit of the sequence in N. Thus N is complete as a metric space. □

As a matter of notation we write $((a_i)_i)_{\mathcal{U}}$ for the image of $(a_i)_{i\in I}$ under the quotient map taking N_0 to N.

Note we can also construct N as done in Section 2.2, by treating M_i as an FOL structure (adding a copy of the reals), taking a FOL ultraproduct and then quotienting.

Proposition 2.3.15. *(Łos' Theorem for continuous logic.) Let M be the ultraproduct of $\{M_i : i \in I\}$ across $\mathcal{U}$ and let $\varphi(x_1, ..., x_n)$ be an $\mathcal{L}$-formula. Let $a_1, ..., a_n \in M$. Let $a^k = ((a_i^k)_{i\in I})_{\mathcal{U}}$ (as in notation above). Then $\varphi(M)(a_1, ..., a_n) = \lim_{i\to\mathcal{U}} \varphi(M_i)(a_i^1, ..., a_i^n)$.*

Proof. The proof is by induction on the structure of the formula φ and by using some basic facts on ultrafilter limits. (See Lemmas 5.1 and 5.2 of [BBHU08].) $\square$

As usual we obtain the compactness theorem.

Proposition 2.3.16. *Let Σ be a set of closed conditions. If Σ is finitely satisfiable then Σ has a model.*

Proof. Let I be the collection of finite subsets of Σ. For each $i \in I$ let M_i be a model of i. For each condition $c \in \Sigma$ let $S(c) = \{i \in I : c \in i\}$. Then the collection of all such subsets $S(c) \subset I$ has the finite intersection property and extends to an ultrafilter $\mathcal{U}$ on I. Let $M = \prod_{\mathcal{U}} M_i$. By Łos's Theorem for CL each $c \in \Sigma$ holds in M. $\square$

Some remarks are in order. Firstly, the above proposition extends to Σ containing free variables with the usual modification of including constants. Secondly, given Σ, let Σ^+ be the set of approximations to conditions Σ, i.e. $\Sigma^+ = \{\sigma \leqslant \frac{1}{n} : n = 1, 2, ..., \text{“}\sigma = 0\text{”} \in \Sigma\}$. Then finite satisfiability of Σ^+ implies that Σ is satisfiable.

We now consider classes of $\mathcal{L}$-structures. We say that a class $\mathcal{C}$ of models is *elementary* if it precisely the class of models of some $\mathcal{L}$-theory.

To give our next proposition which characterizes elementary classes, we must first define the notion of the ultraroot and ultrapower of a structure. Given a structure N and an index set I with ultrafilter $\mathcal{U}$, suppose that $N = \prod_{\mathcal{U}} M_i$ where each $M_i = M$ for some $\mathcal{L}$-structure M. Then N is said to be the *ultrapower* of M along U; N is often denoted $M^I/\mathcal{U}$. In this case, M is said to be the *ultraroot* of N along $\mathcal{U}$.

Proposition 2.3.17. *A nonempty class $\mathcal{C}$ of $\mathcal{L}$-structures is elementary if and only it is closed under isomorphism, ultraproducts, and ultraroots.*

Proof. The forward definition of this proposition is an obvious application of Łoś's Theorem; we prove the reverse direction. So suppose $\mathcal{C}$ is a class of $\mathcal{L}$-structures closed under isomorphism, ultraproducts, and ultraroots. We show $\mathcal{C}$ is elementary. The proof is again routine, but with a slight twist as we do not have negations.

Let $\Sigma = \text{Th}(\mathcal{C})$, that is to say, the set of conditions true in every element of $\mathcal{C}$. By definition, $M \in \mathcal{C}$ implies that $M \models \Sigma$; we show that if $M \models \Sigma$ then $M \in \mathcal{C}$.

For $M \models \Sigma$, let

$$\mathrm{Th}(M)^+ = \left\{ \text{“}\sigma \leqslant \frac{1}{n}\text{”} : \text{“}\sigma = 0\text{”} \in \mathrm{Th}(M) \right\}$$

the set of all approximations to conditions in $Th(M)$ (themselves of course conditions).

We claim that $\mathrm{Th}(M)^+$ is finitely satisfiable in $\mathcal{C}$, namely every finite subset of $Th(M)^+$ has a model in $\mathcal{C}$. If not then clearly for each $\sigma_1, ..., \sigma_n$ such that $\sigma_i = 0$ is true in M, there is $\epsilon > 0$ such that every $N \in \mathcal{C}$ satisfies $\sigma_i \geqslant \epsilon$ for some $i = 1, ..., n$, whereby the condition $max(\sigma_1, ..., \sigma_n) \geqslant \epsilon$ is true in every $N \in \mathcal{C}$ so is in Σ. This contradicts M being a model of Σ, and proves the claim.

Since this set of conditions is finitely satisfiable, the proof of the compactness theorem implies that there is an ultraproduct of elements of $\mathcal{C}$, call it N, which is a model of $\mathrm{Th}(M)^+$. It follows that $N \models Th(M)$ and so $N \equiv M$. Moreover, since $\mathcal{C}$ is closed under ultraproducts, $N \in \mathcal{C}$. (Call this stage in the proof (*), it will be remarked upon below).

A continuous model theory analogue of the Keisler–Shelah theorem implies that since M, N are elementarily equivalent, there is an index set I and ultrafilter $\mathcal{U}$ on I such that $M^I/\mathcal{U} \cong N^I/\mathcal{U}$. Since $\mathcal{C}$ is closed under isomorphisms, $M^I/\mathcal{U} \in \mathcal{C}$. Since it is closed under ultraroots, we have that $M \in \mathcal{C}$, as desired. This concludes the proof. □

Note that another version of this proposition simply replaces the condition that $\mathcal{C}$ is “closed under isomorphism and ultraroots” with the condition that $\mathcal{C}$ is “closed under elementary equivalence”. Note that closure under ultraproducts is still in both statements as well. In order to prove this variant, we simply repeat the same proof but conclude it at stage (*). At that point we had shown that $M \equiv N$ and $N \in \mathcal{C}$, hence we get $M \in \mathcal{C}$.

We also remark that here, as in FOL, closure under elementary equivalence is generally insufficient to guarantee that a class is elementary. This failure is witnessed by classes like the finite fields, or any other class of arbitrarily large finite models of a theory.

We now briefly go over how other basic model theory results generalise to the continuous logic setting.

First unions of chains. There is a natural notion of a chain of $\mathcal{L}$-structures $(M_i : i < \omega)$ for which $i < j$ implies $M_i \subseteq M_j$, where “inclusion” here means substructure. The union of this chain is naturally an $\mathcal{L}$-prestructure with a metric (not a pseudometric). We call its completion $\bigcup_{i \in \omega} M_i$, which is an $\mathcal{L}$-structure. Note that there was nothing special about ω here; we could replace it by any totally ordered set Λ so that $M_\mu \subseteq M_\lambda$ if $\mu < \lambda$ for $\mu, \lambda \in \Lambda$.

Proposition 2.3.18. *If* $(M_\lambda \colon \lambda \in \Lambda)$ *is an elementary chain (i.e.* $M_\mu \preceq M_\lambda$ *if* $\mu < \lambda$*), then for each* $\mu \in \Lambda$, $M_\mu \preceq \bigcup_{\lambda \in \Lambda} M_\lambda$.

Proof. By Tarski–Vaught (Lemma 2.3.13) and induction on formulas. □

We now discuss cardinality considerations in the continuous framework. First, we have called the cardinality $\mathcal{L}$ of our language $\mathcal{L}$, the cardinality of the set of nonlogical symbols (constants, predicates, function symbols) plus ω; hence $|\mathcal{L}|$ is always infinite. Note that assuming we are working with a countable, full set of connectives, $|\mathcal{L}|$ is also the number of $\mathcal{L}$-formulas.

In place of the cardinality of an $\mathcal{L}$-structure M it is better to define $||M||$ as the density character of M, namely the least cardinality of a dense subset. (Because of the stipulation that structures are taken to be complete with respect to the underlying metric.)

Thus, for example, we call a theory T *separably categorical* if all of its separable (i.e. density character is $\aleph_0$) models are isomorphic.

We can now state the continuous analogues of the Lowenheim–Skolem Theorems:

Downward: If M is an $\mathcal{L}$-structure, $\kappa \geqslant |\mathcal{L}|$ a cardinal, and $A \subseteq M$ with $|A| \leqslant \kappa$, then there is $N \preceq M$ with $A \subseteq N$ and $||N|| = \kappa$. This result is proved in the normal way using Tarski–Vaught, simply taking the completion of the resulting prestructure to complete the proof.

Upward: For all $\mathcal{L}$-structures M and cardinals κ, there is $M \preceq N$ such that $||N|| \geqslant \kappa$. Again, this is proved similar to the standard FOL proof.

We now discuss saturation. Note that as in FOL, we denote by $\mathcal{L}_A$ the language $L \cup \{$constants for elements of $A\}$ for a subset A of an $\mathcal{L}$-structure M.

Definition 2.3.19. An $\mathcal{L}$-structure M is *κ-saturated* for a cardinal κ if, whenever $A \subseteq M$ has cardinality $< \kappa$ and $\Sigma(x_1, \cdots, x_n)$ is a set of conditions over A (i.e. in $\mathcal{L}_A$) which is finitely satisfiable in M, then Σ is satisfiable in M.

Proposition 2.3.20. *For every* $\kappa \geqslant \omega$, *any* $\mathcal{L}$*-structure has a* κ*-saturated elementary extension.*

The proof for this proposition is, again, precisely analogous to the ordinary FOL proof.

In continuous model theory, ω-saturated models have the interesting property that we may systematically replace inf and sup quantifiers by actual $\exists$ and $\forall$ quantifiers, made precise in the following proposition.

Proposition 2.3.21. *Suppose we have a formula $\psi(\bar{x})$ which is of the form $Q^1_{y_1} \cdots Q^n_{y_n} \varphi(\bar{x}, \bar{y})$ where the Q^i denote* inf *or* sup *quantifiers and φ is quantifier-free. Let $\tilde{Q^i}$ be $\exists$ if Q^i is* inf, *and $\forall$ if Q^i is* sup. *Suppose M is a ω-saturated $\mathcal{L}$-structure. Then in M, the condition "$\psi(\bar{x}) = 0$" holds of tuple $\bar{a}$ if and only if the "formula" $\tilde{Q^1}_{y_1} \cdots \tilde{Q^n}_{y_n}$("$\varphi(\bar{a}, \bar{y}) = 0$") is true in M.*

Proof. The proof is by induction on the number of quantifiers. We give an example of the key point:

Suppose $\psi(\bar{x}) = \inf_y \varphi(\bar{x}, y)$. In this case, the proposition says that for a ω-saturated model M and $\bar{a} \in M$, $M \models$ "$\psi(\bar{a}) = 0$" if and only if there is $b \in M$ such that $M \models$ "$\varphi(\bar{a}, b) = 0$". Clearly if there is such a b then $M \models \psi(\bar{a})$. To show the other implication, consider the set of conditions $\{$"$\varphi(\bar{a}, y) \leqslant \frac{1}{n}$"$: n = 1, 2, \cdots\}$. Since $M \models (\inf_y \varphi(\bar{a}, y)) = 0$, this set of conditions is finitely satisfiable in M. Then by saturation of M, there is $b \in M$ with $M \models$ "$\varphi(\bar{a}, b) \leqslant \frac{1}{n}$" for all $n \in \omega$. But then $\varphi(\bar{a}, b) = 0$ holds in M, as required. $\square$

This result actually has an analogue in FOL. Suppose $\Sigma(\bar{x}, y)$ is a partial type. The expression $\exists y(\Sigma(\bar{x}, y))$ is *not* a formula but can be viewed as another partial type consisting of its approximations $\exists y(\wedge \Sigma'(\bar{x}, y))$ as Σ' ranges over finite subsets of Σ. Then in an ω-saturated model M, this partial type holds of $\bar{a}$ if there is b such that $M \models \Sigma(\bar{a}, b)$.

A *partial elementary map* $f\colon A \to B$ for $A, B \subseteq M$ is a map such that, for each $\mathcal{L}$-formula $\varphi(\bar{x})$ and $\bar{a} \in A$, $\varphi(M)(\bar{a}) = \varphi(M)(f(\bar{a}))$. A model M is said to be *strongly κ-homogeneous* if, whenever $A, B \subseteq M$ with $|A|, |B| < \kappa$ and $f\colon A \to B$ is a partial elementary map, then f extends to an automorphism of M. As in FOL (and by an analogous proof), we have the existence of κ-saturated and strongly κ-homogeneous models for every κ:

Proposition 2.3.22. *For each model M and $\kappa \geqslant \omega$, M has an elementary extension N which is κ-saturated and strongly κ-homogeneous.*

Given an $\mathcal{L}$-structure M and $\bar{a} \in M^n$, by $\mathrm{tp}_M(\bar{a})$ we mean the set of conditions "$\varphi(\bar{x}) = 0$" satisfied by $\bar{a}$ in M. We can likewise consider $\mathrm{tp}_M(\bar{a}/B)$ which allows for formulas with parameters in B, as usual. Also, let $\mathrm{qftp}_M(\bar{a})$ be the type consisting of all conditions "$\varphi(\bar{x}) = 0$" where φ is a quantifier-free formula.

$tp_M(\bar{a})$ could equivalently be defined as the information giving the value of $\varphi(\bar{a})$ in M for each formula $\varphi(\bar{x})$. Likewise for $qftp_M(\bar{a})$.

Definition 2.3.23. Let T be an $\mathcal{L}$-theory which has a model. We say that T has *quantifier elimination* (QE) if, for every $\mathcal{L}$-formula $\varphi(\bar{x})$ and $\epsilon > 0$, there is a quantifier-free $\mathcal{L}$-formula $\psi(\bar{x})$ such that for any model $M \models T$ and $\bar{a} \in M^n$, $|\varphi(M)(\bar{a}) - \psi(M)(\bar{a})| < \epsilon$.

Lemma 2.3.24. *For an $\mathcal{L}$-theory T, the following are equivalent:*

1. *T has quantifier elimination.*
2. *For any models $M, N \models T$ and tuples $\bar{a} \in M^n$ and $\bar{b} \in N^n$, if $\mathrm{qftp}_M(\bar{a}) = \mathrm{qftp}_N(\bar{b})$ then $\mathrm{tp}_M(\bar{a}) = \mathrm{tp}_N(\bar{b})$.*
3. *Suppose M, N are ω-saturated models of T and $\bar{a} \in M^n$, $\bar{b} \in N^n$ with $\mathrm{qftp}_M(\bar{a}) = \mathrm{qftp}_N(\bar{b})$. Then for any $c \in M$, there is $d \in N$ such that $(\bar{a}, c)$ and $(\bar{b}, d)$ have the same quantifier-free types in their respective models.*

Proof. 1 implies 2 is basically immediate.

To show 2 implies 3, we let M, N be ω-saturated models of T. Take $\overline{a}, \overline{b}$ n-tuples in M and N, respectively, and let $c \in M$. We have to find $d \in N$ such that $\overline{a}c$ and $\overline{b}d$ have the same quantifier-free type. Now, by 2, $\mathrm{tp}_M(\overline{a}) = \mathrm{tp}_N(\overline{b})$. Let $\varphi(\overline{x}, y)$ be a quantifier-free formula such that $M \models \varphi(\overline{a}, c) = 0$. Then $M \models \inf_y \varphi(\overline{a}, y) = 0$, and so $N \models \inf_y \varphi(\overline{b}, y) = 0$ as well. Thus, by ω-saturation of N, there is $d \in N$ such that $N \models \inf_y \varphi(\overline{b}, d) = 0$. This extends to finitely many formulas by considering the *max* connective. Use the ω-saturation of N again to find d such that $N \models \varphi(\overline{b}, d) = 0$ for *all* φ quantifier-free such that "$\varphi(\overline{x}, y) = 0$" $\in \mathrm{tp}_M(\overline{a}c)$.

3 implies 2: If $M, N \models T$ $\overline{a} \in M^n$, $\overline{b} \in N^n$, and $\mathrm{qftp}_M(\overline{a}) = \mathrm{qftp}_N(\overline{b})$, we show by induction on the complexity of φ that $M \models \varphi(\overline{a}) = 0$ if and only if $N \models \varphi(\overline{b}) = 0$. Let us consider the special case where $\varphi(\overline{x}) = \inf_y \psi(\overline{x}, y)$, where $\psi(\bar{x}, y)$ is quantifier-free. We may assume (by passing to elementary extensions) that M and N are ω-saturated. Suppose $\varphi(M)(\bar{a}) = 0$, so by ω-saturation there is $c \in M$ with $\psi(M)(\bar{a}, c) = 0$. By 3, there is $d \in N$ such that $\bar{a}, c)$ and $(\bar{b}, d)$ have the same quantifier-free type in M, N respectively. Hence $\psi(N)(\bar{b}, d) = 0$, whereby $\varphi(N)(\bar{b}) = 0$.

Finally, for 2 implies 1 we need to know something about the topologies on the type space $S_n(T)$ (the collection of complete n-types of T). This will be discussed in detail in Section 2.3.4. We get two topologies on $S_n(T)$, τ_1 where the basic opens are given by formulas (or conditions), and τ_2 where there basic opens are given by quantifier-free formulas. Consider the identity map, $i : S_n(T) \to S_n(T)$. It is clearly a continuous map from $(S_n(T), \tau_1)$ to $(S_n(T), \tau_2)$ (as a quantifier-free formula is a formula). Now τ_1 is a compact Hausdorff topology. By assumption 2, so is τ_2. A basic topological fact is that a continuous bijection between compact Hausdorff spaces is a homeomorphism. Hence the two topologies coincide, and a further little argument gives 1. □

Before passing on to an example, let us mention some trivialities about *sup*, *inf* and $\forall$.

Remark 2.3.25. Let $\varphi(\bar{x}, \bar{y})$ be a formula, and let $\psi(\bar{x})$ be $\sup_{\bar{y}} \varphi$. Then for any $\mathcal{L}$-structure M and $\bar{a}$ from M the condition $\psi = 0$ is true of $\bar{a}$ in M if and only if for all $\bar{b} \in M$, $\varphi(\bar{a}, \bar{b}) = 0$. So we can use $\forall \bar{y}(\varphi(\bar{x}, \bar{y}) = 0)$ as alternative notation for $\psi(\bar{x}) = 0$.

Likewise we can write $\forall \bar{y}(\varphi \leqslant r)$ for $\psi \leqslant r$ and $\forall \bar{y}(\varphi \geqslant r)$ for $(inf_{\bar{y}}\varphi) \geqslant r$.

In any case, by a *universal* condition we will mean one of the form $(sup_{\bar{y}}\varphi) = 0$ for $\varphi(\bar{x}, \bar{y})$ a quantifier-free formula.

Example 2.3.26 (Atomless probability algebras). We want to give an "interesting" example of a theory with quantifier elimination. Note that any FOL theory with quantifier elimination is, in particular, a CL theory with quantifier elimination. However, the issue is to find *strictly* CL theories with QE. Atomless probability algebras will be such an example. (Note that we have already briefly discussed probability algebras in CL in Example 2.3.2.) Atomless probability algebras are very well studied structures we will treat them following [BU10].

Recall that a (abstract) Boolean algebra is a set B together with distinguished elements 0 and 1, binary operations $\wedge$, $\vee$, and a unary operation ^{c} satisfying the following universal axioms:

(i) Associativity for $\wedge$ and $\vee$:

$$\forall x \forall y \forall z((x \wedge y) \wedge z = x \wedge (y \wedge z))$$
$$\forall x \forall y \forall z((x \vee y) \vee z = x \vee (y \vee z))$$

(ii) Commutativity for $\wedge$ and $\vee$:

$$\forall x \forall y(x \wedge y = y \wedge x)$$
$$\forall x \forall y(x \vee y = y \vee x)$$

(iii) Distributivity of $\wedge$ over $\vee$ and vice versa:

$$\forall x \forall y \forall z (x \wedge (y \vee z) = (x \wedge y) \vee (x \wedge z))$$
$$\forall x \forall y \forall z (x \vee (y \wedge z) = (x \vee y) \wedge (x \vee z))$$

(iv) Absorption axioms:

$$\forall x \forall y (x \vee (x \wedge y) = x)$$
$$\forall x \forall y (x \wedge (x \vee y) = x)$$

(v) $\forall x (x \vee 0 = x)$ and $\forall x (x \wedge 1 = x)$
(vi) $\forall x (x \vee x^{\mathrm{c}} = 1)$ and $\forall x (x \wedge x^{\mathrm{c}} = 0)$.

Note. We could define $x \leqslant y$ to be $x \vee y = x$. In that case $\leqslant$ will be a partial ordering and we can formulate all the axioms in the language of partial orderings.

A typical example of a Boolean algebra is a *concrete* Boolean algebra, namely a collection of subsets of a given set Ω which contains $\varnothing$, Ω, and is closed under union, intersections, and complementation.

Any (abstract) Boolean algebra is isomorphic to a concrete Boolean algebra. A typical example of a concrete Boolean algebra is the collection of definable subsets of a FOL structure M.

What about probability algebras as CL-structures? Well, we have a Boolean algebra B with a $[0,1]$-valued metric d, and $[0,1]$-valued formula μ. We can rewrite the axioms of Boolean algebras in terms of d, bearing in mind earlier conventions about universal quantifiers. For example, $\forall x \forall y (d(x \wedge y, y \wedge x) = 0)$ expresses commutativity of $\wedge$. The formula μ, satisfies $\mu(1) = 1$, $\mu(0) = 0$ and finite additivity:

$$\forall x \forall y (\mu(x \vee y) = \mu(x \wedge y^{\mathrm{c}}) + \mu(y)).$$

From this it follows that $\mu(x) \leqslant \mu(y)$ whenever $x \leqslant y$ (as defined in the note above). Finally, we have an additional statement relating d and μ:

$$\forall x \forall y (d(x, y) = \mu((x \wedge y^{\mathrm{c}}) \vee (x^{\mathrm{c}} \wedge y))$$

saying that the distance between x and y is the measure of their symmetric difference.

Thus, we are working with the language $L = \{0, 1, \wedge, \vee, ^{\mathrm{c}}, \mu\}$ (and d as a "logical symbol"). Also note that, if M is a model of the axioms, then $\mu(M)(a) = 0$ implies $d(a, 0^M) = 0$ and so $a = 0$.

An *atomless probability measure* is just an probability algebra such that the underlying Boolean algebra is atomless, where an *atom* is an element a

such that there exists no element b in the algebra that satisfies $0 < b < a$. In [BU10], atomlessness is expressed by the condition

$$\sup_x \inf_y \left| \mu(y \wedge x) - \frac{1}{2}\mu(x) \right| = 0.$$

We have to check that any L-pre-structure satisfying the axioms is actually complete. This is done in [BBHU08] Section 16.

Let us now discuss how/why T has quantifier elimination. We begin by talking about QE for atomless Boolean algebras in FOL. First note that in atomless Boolean algebras, for any $a \neq 0$, there are nonzero disjoint $c, d \leqslant a$. Let M, N be models. Let $\overline{a} = (a_0, \ldots, a_{n-1}) \in M^n$ and $\overline{b} = (b_0, \ldots, b_{n-1}) \in N^n$ with the same quantifier-free type. This implies that the map $a_i \mapsto b_i$ extends to a isomorphism between $\langle \overline{a} \rangle$ and $\langle \overline{b} \rangle$. For each $I \subseteq n$ and let

$$\overline{a}_I = \bigwedge_{i \in I} a_i \wedge \bigwedge_{j \notin I} a_j^c,$$

and likewise for $\overline{b}_I$. Then the $\overline{a}_I$ are pairwise disjoint and $\overline{a}_I = 0$ if and only if $\overline{b}_I = 0$.

Now, let $c \in M$. Consider $\{c \wedge \overline{a}_I : I \subseteq n\} \cup \{c^c \wedge \overline{a}_I : I \subseteq n\}$. By atomlessness, there is $d \in N$ such that $d \wedge \overline{b}_I = \varnothing$ if and only if $c \wedge \overline{a}_I = \varnothing$, $d^c \wedge \overline{b}_I = \varnothing$ if and only if $c \wedge \overline{a}_I = \varnothing$. Then d works.

Now let us modify the construction in the setting of atomless probability algebra. With the earlier notation, we have in addition the quantifier-free information $\mu(\overline{a}_I) = \mu(\overline{b}_I)$ for each $I \subseteq \{1, ..., n\}$. Given $c \in M$, use the axioms again to find $d \in N$ such that $\mu(d \wedge \overline{b}_I) = \mu(c \wedge \overline{a}_I)$ and $\mu(d^c \wedge \overline{b}_I) = \mu(c^c \wedge \overline{a}_I)$ for all I, and d works.

By similar means and analogous to the FOL case we can show that theory of atomless probability algebras is separably categorical by doing a back and forth argument in separable models and taking the completion. We obtain completeness of the theory by the CL analogue of Vaught's test.

2.3.4 *Type spaces and definability*

As motivation, let us first recall the situation in FOL. Let T be a complete theory in a, say 1-sorted, FOL language L. Let $S_n(T)$ be the "space" of complete n-types of T, i.e. $\{\text{tp}_M(\overline{a}) : M \models T, \overline{a} \in M^n\}$. We topologize $S_n(T)$ by taking as "basic closed" sets, sets of the form $[\varphi(\overline{x})] = \{p(\overline{x}) : \varphi \in p\}$, where $\varphi(\overline{x})$ is an L-formula. With this definition, and taking closed subset of $S_n(T)$ to be arbitrary intersections of basic closed sets, we get a topology on $S_n(T)$. Note that, given φ, $[\neg\varphi]$ is exactly the complement of $[\varphi]$ in $S_n(T)$. From this it follows that the basic closed sets are also open, so they are in fact clopen. Thus, we have a basis of clopen sets. $S_n(T)$ is also Hausdorff and compact. This is the so called "Stone topology" on $S_n(T)$ the space of ultrafilters on the Boolean algebra of L-formulas $\varphi(\overline{x})$ up to equivalence mod T. To repeat, $S_n(T)$ is a compact Hausdorff space with a basis of clopen sets, also called a compact totally disconnected space, also called a profinite space.

Fact. *The clopen subsets of $S_n(T)$ are precisely the $[\varphi]$ for $\varphi(\overline{x})$ a formula.*

Corollary. The continuous functions $f : S_n(T) \to \{0,1\}$ correspond to formulas $\varphi(\overline{x})$ modulo T.

This is because $f^{-1}(0)$ is a clopen subset of $S_n(T)$, so equal to $\{p(\overline{x}) : \varphi \in p\}$ for some $\varphi(\overline{x})$, and so $f(p) = 0$ if and only if $\varphi \in p$.

What do arbitrary non-empty closed subsets $C \subseteq S_n(T)$ correspond to? Well, in that case $C = \bigcap_{\varphi \in \Phi(\overline{x})} [\varphi(\overline{x})]$, and C corresponds to the partial type $\Phi(\overline{x})$ of T. The set that such a type defines in a model is called a type-definable set, or a $\bigwedge$-definable set, or a ∞-definable set. And what about arbitrary open sets $U \subseteq S_n(T)$? In this case we have $U = \bigcup_{\varphi \in \Phi(\overline{x})} [\varphi(\overline{x})]$, so U corresponds to $\bigvee \Phi(\overline{x})$. Such a set, when interpreted in a model, is called a $\bigvee$-definable set, or an ind-definable set.

As an aside, let us note the following. Given T, we associate $S_n(T)$ to $n = \{0, \ldots, n-1\}$. Now, given a function $f : n \to m$, we get a continuous function $S(f) : S_m(T) \to S_n(T)$ by defining $S(f)(\text{tp}(a_0, \ldots, a_{m-1})) = \text{tp}(a_{f(0)}, \ldots, a_{f(n)})$. This yields a contravariant functor (the type-space functor) from which we can recover the entire theory. Note that when f is the inclusion function $n \to n+1$, then $S(f)$ s not only continuous but also open, as the image of the clopen set given by the formula $\varphi(x_0, ..., x_n)$ is the clopen set given by $\exists x_n \varphi$.

Now, what happens in continuous logic? In this case, it is convenient again to work with a complete CL-theory T in language L, so $T = \text{Th}(M)$ for some M, as discussed earlier (the collection of sentential conditions $\sigma = 0$ true in M, equivalent the set of pairs $(\sigma, \sigma(M))$ for σ an L-sentence).

Definition 2.3.27. $S_n(T)$ is the "space" $\{\text{tp}_M(\overline{a}) : M \models T, \bar{a} \in M^n\}$ of complete n-types of T.

Definition 2.3.28. By a basic closed set in $S_n(T)$ we mean $\{p(\overline{x}) \in S_n(T) :$ "$\varphi(\overline{x}) = 0$" $\in p(\overline{x})\}$ for some L-formula $\varphi(\overline{x})$.

Note that, if we assume all continuous functions $[0,1] \to [0,1]$ to be connectives, then for any closed subset $D \subseteq [0,1]$ the expression "$\varphi(\overline{x}) \in D$" is equivalent to some condition $\psi(\overline{x}) = 0$. This holds because there is a continuous function $u : [0,1] \to [0,1]$ such that $D = u^{-1}(0)$, so $u(\varphi(\overline{x})) = 0$ if and only if $\varphi(\overline{x}) \in D$.

Lemma 2.3.29. *Defining closed subsets of $S_n(T)$ to be arbitrary intersections of basic closed sets equips $S_n(T)$ with a compact Hausdorff topology.*

Proof. To see that $S_n(T)$ is a topological space: The whole space is closed via the condition "$d(x,x) = 0$". The empty set is closed via the condition "$1 = 0$". Hence both the whole space and empty set are open. By definition, a closed set is the intersection of basic closed sets, so the arbitrary intersection of closed sets is closed.

To see that $S_n(T)$ is Hausdorff, let $p(\bar{x}) \neq q(\bar{x}) \in S_n(T)$. We must show that p and q are separated by open disjoint sets. Since p and q are distinct, there is some $\varphi(\bar{x})$ such that $\varphi(p) = r \neq s = \varphi(q)$. (Here $\varphi(p) = \varphi(M)(\bar{a})$ where $\bar{a} \models p(\bar{x})$ in $M \models T$.) Let $I, J \subseteq [0,1]$ be open disjoint intervals such that $r \in I$ and $s \in J$. Then the expressions "$\varphi(\bar{x}) \in I$" and "$\varphi(\bar{x}) \in J$" define open disjoint sets in $S_n(T)$ containing p and q respectively.

To see that $S_n(T)$ is compact, suppose C_i, $i \in I$ are closed sets with the finite intersection property, i.e. any finite intersection of the C_i is nonempty. We must show that $\bigcap_I C_i$ is nonempty. We may assume each C_i is basic closed, and given by the condition "$\varphi_i(\bar{x}) = 0$". Then theory $T \cup \{$"$\varphi_i(\bar{x}) = 0$"$: i \in I\}$ is finitely satisfiable. So by the compactness theorem, it has a model, i.e. there is $M \models T$ and $\bar{a} \in M$ such that $\varphi_i(M)(\bar{a}) = 0$ for all $i \in I$. Then $p = \mathrm{tp}_M(\bar{a}) \in \bigcap_I C_i$. □

Note that as for FOL, we get a contravariant functor $S : \mathbb{N} \to \mathrm{Top}$ via $S(n) = S_n(T)$ in continuous logic. We leave it as an exercise to show that if $f : n \to m$ is one-one then $S(f)$ is surjective and open (in addition to being continuous).

For $\varphi(\bar{x})$ and $p \in S_n(T)$, define $\varphi(p) = \varphi(M)(\bar{a}) \in [0,1]$ for $\bar{a} \in M \models T$ realizing p. With this notation:

Lemma 2.3.30. $\varphi : S_n(T) \to [0,1]$ *is continuous.*

Proof. We must show that the preimage of a closed set $D \subseteq [0,1]$ is closed.

Indeed, $\varphi^{-1}(D) = \{p(\bar{x}) \in S_n(T) : \varphi(p) \in D\}$, which is given by the condition "$\varphi(\bar{x}) \in D$". □

Let $\Sigma(\bar{x}) = \{$"$\varphi_i(\bar{x}) = 0$" $: i \in I\}$. This corresponds to a closed set in $S_n(T)$, the CL version of a "partial type":

Definition 2.3.31. Fix complete T. By a (partial) type $\Sigma(\bar{x})$ over $\varnothing$ (i.e., without parameters), we mean a set of conditions "$\varphi(\bar{x}) = 0$".

Given $M \models T$, the solution set $\Sigma(M)$ of $\Sigma(\bar{x})$ in M is $\{\bar{a} \in M^n : \varphi(M)(\bar{a}) = 0$ for each condition "$\varphi(\bar{x}) = 0$" $\in \Sigma(\bar{x})\}$.

We will assume that a partial type Σ is closed under "finite conjunctions" (as discussed above) as well as "implications". The latter means that if a condition holds of all solutions of Σ in all models of T then the condition is in Σ.

We say two partial types $\Sigma(\bar{x}), \Gamma(\bar{x})$ are equivalent (mod T) if $\Sigma(M) = \Gamma(M)$ for every $M \models T$.

Lemma 2.3.32. *Closed subsets of $S_n(T)$ correspond to (are in bijection with) partial types $\Sigma(\bar{x})$ up to equivalence mod T.*

Proof. A closed subset of $S_n(T)$ is by definition an intersection of basic closed sets each given by some condition $\varphi(\bar{x}) = 0$, and this set of conditions is precisely a partial type.

One has to check that two partial types are equivalent iff they define the same closed subset of $S_n(T)$. Suppose first that Σ and Γ are equivalent. Let p be a complete type in the closed set defined by Σ. Let M be a model of T in which p is realized. By definition of equivalence p is a complete type in the closed set determined by Γ. The converse is similar: if Σ and Γ are not equivalent, let M be a model of T in which $\Sigma(M) \neq \Gamma(M)$. Without loss $\bar{a} \in \Sigma(M) \backslash \Gamma(M)$. Let $p = tp(\bar{a})$. By assumption there is a condition $\varphi(\bar{x}) = 0$ which is in Γ but does not hold of $\bar{a}$, so $\varphi(\bar{a}) = \epsilon > 0$ for some ϵ. Hence p is not in the closed set determined by Γ, whereas it is in the closed set determined by Σ. □

In the FOL case we have seen that formulas correspond to continuous functions from $S_n(T)$ to $\{0, 1\}$. The CL case is a bit more complicated and will be discussed in Section 2.3.4.1.

We finish this part with a discussion of the natural metric on type spaces. Recall first that for $\bar{a}, \bar{b} \in M^n$, $M \models T$, $d(M)(\bar{a}, \bar{b}) = \max\{d(M)(a_i, b_j)\}$.

Definition 2.3.33. For $p, q \in S_n(T)$, define

$$d(p,q) = \inf\{d(M)(\bar{a},\bar{b}) : \bar{a},\bar{b} \in M^n,\ M \models T,\ \bar{a} \models p,\ \bar{b} \models q\}.$$

It is pretty clear that d defines a metric on $S_n(T)$. Suppose for example that $d(p,q) = 0$. Let $\bar{a}$ realize p in a saturated model M. For each $\epsilon > 0$ there is a realization $\bar{b}$ of q with $d(\bar{a},\bar{b}) < \epsilon$. By saturation there is $\bar{b}$ in M realizing q with $d(\bar{a},\bar{b}) = 0$ namely $\bar{a} = \bar{b}$ so $q = p$.

Lemma 2.3.34.

(i) The d-metric topology on $S_n(T)$ refines the type-space or "logic" topology.

(ii) For any closed $F \subseteq S_n(T)$ in the logic topology, and for any $\epsilon > 0$, the ϵ-neighborhood of F, i.e. the set $\{q \in S_n(T) : \exists p \in F : d(p,q) \leqslant \epsilon\}$, is closed in the logic topology.

(iii) $(S_n(T), d)$ is a complete metric space (although not necessarily compact).

(iv) For any $\mathcal{L}$-formula $\varphi(\bar{x})$, the map $\varphi : S_n(T) \to [0,1]$ is uniformly continuous in the d-metric topology.

Proof.

(i) Let $p(\bar{x}) \in S_n(T)$, and suppose $\varphi(p) = r$. Fix an open neighborhood $I = (r - \epsilon, r + \epsilon)$ of r in $[0,1]$. Then the "open condition" "$\varphi(\bar{x}) \in I$" defines a basic open neighborhood of p in the logic topology. Now fix $\bar{a} \models p$, in a saturated model $M \models T$. Then by uniform continuity of the function $\varphi : M \to [0,1]$, there is $\delta > 0$ such that $d(M)(\bar{a},\bar{b}) < \delta \Rightarrow \varphi(M)(\bar{b}) \in I$. It follows that $d(p,q) < \delta \Rightarrow \varphi(q) \in I$. Thus, the logic-open set ["$\varphi(\bar{x}) \in I$"] contains an open neighborhood of p in the d-metric topology, hence it is open in the d-metric topology.

(ii) F being closed in the logic topology means that F is given by a partial type $\Sigma(\bar{x})$. The closed ϵ-neighborhoods of F in $S_n(T)$ are given by

$$(\exists \bar{y})(d(\bar{x},\bar{y}) \leqslant \epsilon \wedge \Sigma(\bar{y})).$$

However, $d(\bar{x},\bar{y}) \leqslant \epsilon \wedge \Sigma(\bar{y})$ defines a logic-closed set $D \subseteq S_{2n}(T)$. So the set in question is the projection of D to $S_n(T)$, so it is closed since the projection map is closed.

(iii) This is a general fact. Namely, given a compact Hausdorff space X and a metric d refining that topology, such that the ϵ-closed neighborhood of any topologically closed set is topologically closed, then (X,d) is complete. The idea is to take a Cauchy sequence and use compactness to find a limit point.

(iv) By (i), $\varphi : S_n(T) \to [0,1]$ is continuous in the d-metric topology. To see that it is uniform, let Δ_φ be a modulus of uniform continuity of φ for T (or equivalently, for $\mathcal{L}$). Recall that this means that $\varphi(M) : M^n \to [0,1]$ has Δ_φ as a modulus for any $M \models T$. Then we claim that Δ_φ is also a modulus for $\varphi : S_n(T) \to [0,1]$.

Given $\epsilon > 0$, let $\delta = \Delta_\varphi(\epsilon)$. Let $d(p,q) < \delta$. Then there exist $M \models T$, and $\bar{a}, \bar{b} \in M^n$ realizing p and q respectively, such that $d(M)(\bar{a}, \bar{b}) < \delta$. This means that $|\varphi(M)(\bar{a}) - \varphi(M)(\bar{b})| < \epsilon$, so $|\varphi(q) - \varphi(p)| < \epsilon$. □

2.3.4.1 *Continuous function on type spaces*

For a continuous logic theory T, we want to consider continuous functions from $S_n(T)$ to $[0,1]$ for some fixed n. The key to describing such functions is a strengthening of the Stone–Weierstrass theorem, which can be found in Proposition 1.4 of [BU10].

Fact 2.3.35. *Let X be a compact Hausdorff space, and consider $\mathcal{U} = C(X,[0,1])$ (continuous functions from X to $[0,1]$) with the uniform convergence topology. Let $\mathcal{B} \subseteq \mathcal{U}$ have the following properties:*

- *If $f \in \mathcal{B}$, then the function $\neg f$ defined via $\neg f(x) = 1 - f(x)$ for all $x \in X$ is also in $\mathcal{B}$.*
- *If $f, g \in \mathcal{B}$, then the function $f \dot{-} g$ defined via $f \dot{-} g(x) = f(x) - g(x)$ if $f(x) > g(x)$ and 0 otherwise for all $x \in X$ is also in $\mathcal{B}$.*
- *If $f \in \mathcal{B}$, then the function $\frac{f}{2}$ defined via $\frac{f}{2}(x) = \frac{f(x)}{2}$ for all $x \in X$ is also in $\mathcal{B}$.*
- *$\mathcal{B}$ separates points, i.e. if $x, y \in X$ with $x \neq y$, there exist $f, g \in \mathcal{B}$ with $f(x) \neq g(y)$.*

Then, $\mathcal{B}$ is dense in $\mathcal{U}$.

Recall that in Section 2.3.2.3, we discussed different collections of propositional connectives. We shall now formally assume that our propositional connectives satisfy the assumptions of Fact 2.3.35 (taking $X = [0,1]^n$).

We say $\Phi : S_n(T) \to [0,1]$ is uniformly approximated by $\mathcal{L}$-formulas if there is a sequence of $\mathcal{L}$-formulas $\varphi_n(\overline{x}) : S_n(T) \to [0,1]$ such that for every $\epsilon > 0$ there exists an N such that for every $n > N$, $|\Phi(p) - \varphi_n(p)| < \epsilon$ for every $p \in S_n(T)$.

Proposition 2.3.36. *The continuous functions from $S_n(T)$ to $[0,1]$ are precisely the functions which are uniformly approximated by $\mathcal{L}$-formulas.*

Proof. For the reverse direction, we know that each $\mathcal{L}$-formula $\varphi(\overline{x})$ corresponds to a continuous function from $S_n(T)$ to $[0,1]$, so it is immediate that any function uniformly approximated by $\mathcal{L}$-formulas is continuous also.

For the forward direction, the collection of $\mathcal{L}$-formulas is closed under $\neg$, $\dot{-}$, and $f/2$ by the assumption on our propositional connectives. Furthermore, it is immediate that the collection of $\mathcal{L}$-formulas separates points in $S_n(T)$ by definition: if two types are not equal, the value assigned to some formula must differ, and therefore that formula separates them. Therefore, the collection of $\mathcal{L}$-formulas is dense in $C(S_n(T),[0,1])$ by Fact 2.3.35. As this space is equipped with the uniform convergence topology, this is precisely saying that every continuous function from $S_n(T)$ to $[0,1]$ is uniformly approximated by $\mathcal{L}$-formulas. □

Definition 2.3.37. Fix a complete $\mathcal{L}$-theory T and some n. By an n-ary definable predicate (with respect to T), we mean a $[0,1]$-valued relation P on models of T induced by a continuous function $\Phi : S_n(T) \to [0,1]$. Namely for $M \models T$ and $\bar{a}$ an n-tuple from M, $P(\bar{a}) = \Phi(tp_M(\bar{a}))$.

Remark 2.3.38.

- Note that if $\mathcal{M}$ is saturated, then we can recover Φ from P, but we cannot do so in general.
- Another way to think of these objects is as "generalized formulas."
- A priori, we should be able to systematically add a new predicate symbol to $\mathcal{L}$ for each definable predicate and add its definition to T. This would give us a new CL language $\mathcal{L}'$ and a complete $\mathcal{L}'$-theory T', which is the CL analogue of Morleyization. The reader is invited to write down an axiomatization of T'.
- We have already alluded to definable predicates in Section 2.2 when discussing C-valued formulas in FOL.

Example 2.3.39. Let T be a complete FOL theory in language L, which we can view as a CL theory (with discrete metric). Let $\varphi_1(\overline{x}), \varphi_2(\overline{x}), \ldots$ be L-formulas. Then for each N, $\psi_N(\overline{x}) = \Sigma_{k=1}^{N} 2^{-k}\varphi_k(\overline{X})$ is a CL-formula which is not $\{0,1\}$-valued. And $\lim_N \psi_N(\overline{x})$ is a definable predicate.

What is the connection between definable predicates and type-definable sets (in continuous logic)?

Remark 2.3.40. Let $\mathcal{M}$ be a saturated model of a complete CL-theory T and let $P : M^n \to [0,1]$. Then P is an n-ary definable predicate if and only if $\{\overline{a} \in M^n : P(\overline{a}) \in D\}$ is type definable for each closed $D \subseteq [0,1]$.

Proof. This is really just a restatement of our definition that P is induced by a continuous function from $S_n(T)$ to $[0,1]$. But we give the proof anyway.

Assume P is an n-ary definable predicate and $D \subseteq [0,1]$ is closed. Then by definition there is a continuous function $\hat{P} : S_n(T) \to [0,1]$ such that $P(\overline{a}) = \hat{P}(tp_{\mathcal{M}}(\overline{a}))$ for all $\overline{a} \in M^n$. Thus $\{\overline{a} \in M^n : P(\overline{a}) \in D\} = \{\overline{a} \in M^n : \hat{P}(tp_{\mathcal{M}}(\overline{a})) \in D\}$. As $\mathcal{M}$ is saturated, $\hat{P}^{-1}(D) = \{tp_{\mathcal{M}}(\overline{a}) : \overline{a} \in M^n$ and $P(\overline{a}) \in D\}$. Furthermore, this set is closed because $\hat{P}$ is continuous and D is closed. However, this means that $\{\overline{a} \in M^n : P(\overline{a}) \in D\}$ is type definable by definition.

Now suppose that $\{\overline{a} \in M^n : P(\overline{a}) \in D\}$ is type-definable for every $D \subseteq [0,1]$. Let $p \in S_n(T)$. As $\mathcal{M}$ is sufficiently saturated, $p = tp_{\mathcal{M}}(\overline{a})$ for some $\overline{a} \in M^n$, so define $\hat{P}(p) = P(\overline{a})$. Then for any closed $D \subseteq [0,1]$, $\hat{P}^{-1}(D) = \{tp_{\mathcal{M}}(\overline{a}) : P(\overline{a}) \in D\}$ is type definable by assumption, and therefore closed as subset of $S_n(T)$. Thus $\hat{P}$ is a continuous function from $S_n(T)$ to $[0,1]$ inducing P, so P is an n-ary definable predicate. □

So far, we have been using definability without parameters, i.e. everything has been definable over $\varnothing$. However, everything we have done can be generalized to work with parameters. This allows us to prove results like the following which are analogs of results in FOL.

Suppose $\mathcal{M}$ is κ-saturated and strongly κ-homogeneous. Let A, B be subsets of M of cardinality $< \kappa$. Suppose $P : M^n \to [0,1]$ is a definable over B predicate which is invariant under automorphisms of $\mathcal{M}$ fixing A pointwise. Then P is a definable over A predicate.

There is a more explicit syntactic characterization of definable predicates given in Proposition 9.3 of [BBHU08]. (Alternatively, one can use the idea of "forced limits" from [BU10].)

Lemma 2.3.41. *The definable n-ary predicates correspond to formal expressions $u(\varphi_1(\overline{x}), \varphi_2(\overline{x}), \dots)$ for formulas $\varphi_k(\overline{x})$, $k \in \omega$, an n-tuple of variable symbols $\overline{x}$, and u a continuous function from $[0,1]^\omega \to [0,1]$.*

Remark 2.3.42. We (they) define a 0-set, or zero set, in a model $\mathcal{M}$ of a CL-theory T to be $\{\overline{a} \in M^n : P(\overline{a}) = 0\}$, where P is an n-ary definable predicate. Notice that this is just a condition generalized to our "generalized formulas." In fact, zero sets are exactly the countably type-definable over $\varnothing$ sets, as we see below.

Lemma 2.3.43. *A set $D \subseteq M^n$ is a zero set if and only if there are countably many $\mathcal{L}$-formulas $\varphi_1(\overline{x}), \varphi_2(\overline{x}), \dots$ such that $D = \bigcap_{k=1}^{\infty} \{\overline{a} \in M^n : \varphi_k(\overline{a}) = 0\}$.*

Proof. For the forward direction, suppose $D = \{\overline{a} \in M^n : P(\overline{a}) = 0\}$, where P is an n-ary definable predicate. By Proposition 2.3.36, there is a sequence of $\mathcal{L}$-formulas $\varphi_k(\overline{x})$ such that $|P(\overline{a}) - \varphi_k(\overline{a})| < \frac{1}{k}$ for all $\overline{a} \in M^n$ and $k \geqslant 1$. Then it is immediate that $D = \bigcap_{k=1}^{\infty}\{\overline{a} \in M^n : (\varphi_k \dot{-} \frac{1}{k})(\overline{a})\}$.

Suppose that $D = \bigcap_{k=1}^{\infty}\{\overline{a} \in M^n : \varphi_k(\overline{a}) = 0\}$ for some sequence of $\mathcal{L}$-formulas $\varphi_k(\overline{x})$. Then for the definable predicate P defined via $P = \Sigma_{k=1}^{\infty} 2^{-k}\varphi_k$, D is clearly the zero set of P. □

Thus far, we have discussed the definable predicates, type-definable sets, and zero sets. Finally, we shall discuss "definable sets". The notion is somewhat obscure and esoteric, but I am told by the experts that it is important. We start with the official definition.

Definition 2.3.44. Let $\mathcal{M}$ be a saturated model of a complete CL-theory T and let $D \subseteq M^n$ be closed in the metric. The we say D is definable (without parameters) if the function $dist(\overline{x}, D) = \inf_{\overline{y} \in D}\{d(\overline{x}, \overline{y})\}$ is a definable predicate.

Remark 2.3.45.

- Such a definable set is also a 0-set. Because let $F(x) = dist(x, D)$ a definable predicate. If $a \in D$ then $F(a) = 0$. Conversely if $F(a) = 0$ then for all n there is $b_n \in D$ such that $d(a, b_n) \leqslant 1/n$. Then $(b_n)_n$ is a Cauchy sequence, so has a limit in M^n, which is in D as D is closed.
- Suppose the metric is discrete. Then any definable set D is the solution of an "FOL-formula", namely there is a $\{0, 1\}$-valued definable predicate F such that D is the 0-set of F. This is because $dist(x, D) = 0$ if $x \in D$ and 1 otherwise.
- From this, we can see that there are zero sets which are not definable: take any FOL set which is type-definable but not definable. Then it will be the zero set of some definable predicate by Lemma 2.3.43, but it will not be definable by the previous remark as it is not definable in the FOL sense. We shall see a nontrivial example of this phenomenon later.
- Suppose $D \subseteq M^n$ is a definable set given by $F(\overline{x}) = dist(\overline{x}, D)$. Let $\mathcal{N}$ be an elementary substructure of $\mathcal{M}$. Then $D \cap N^n$ is the zero set of F restricted to N, i.e. $dist(\overline{x}, D \cap N^n)$ computed inside of $\mathcal{N}$. This follows from the fact that $\mathcal{N}$ will be complete and the definition of elementary substructures.

Proposition 2.3.46. *Let $\mathcal{M}$ be a saturated model of a CL-theory T and suppose that $D \subseteq M^n$ is closed in the metric. Then the following are equivalent:*

- *D is a definable set (without parameters).*
- *The collection of definable predicates is closed under restricting the quantifiers to D. That is, if $P : M^{k+n} \to [0,1]$ is a definable predicate, then so are the functions sending $\overline{x}$ to $\inf_{\overline{y} \in D}\{P(\overline{x}, \overline{y}\}$ and $\sup_{\overline{y} \in D}\{P(\overline{x}, \overline{y})\}$.*

Proof. If the definable predicates are closed under restricting the quantifiers to D, then as $d(\overline{x}, \overline{y}) : M^{2n} \to [0,1]$ is a definable predicate, so is the function sending $\overline{x}$ to $\inf_{\overline{y} \in D}\{d(\overline{x}, \overline{y})\}$. By definition, then, D is definable.

For the other direction, see [BBHU08] Theorem 9.17. □

Remark 2.3.47. Let M be an FOL structure and $\varphi(\bar{x})$ an L-formula. A way of viewing $\varphi(M)$ as an FOL structure was described in Section 2.2. In an analogous manner one can give a definable set D in CL an induced structure.

Finally in this section, we discuss an example pointed out to us by Ward Henson.

Consider the theory of pointed ultrametric spaces of diameter 1. The language L has only a constant symbol c in addition to the metric d. The usual metric space axiom $d(x,z) \leqslant d(x,y) + d(y,z)$ is strengthened to (*) $d(x,z) \leqslant max\{d(x,y), d(y,z)\}$.

We also require that $d(x,c) \leqslant 1$ holds for all x and that for some x, $d(x,c) = 1$.

Let us discuss the meaning of the ultrametric axiom $(*)$?

1) $\forall r \in [0,1]$, $d(x,y) \leqslant r$ and $d(x,y) < r$ both define equivalence relations. So of course $d(x,y) < r$ refines $d(x,y) \leqslant r$ and for $r \leqslant s$ $d(x,y) \leqslant r$ refines $d(x,y) \leqslant s$.
2) By a closed ball of radius r we mean an equivalence class of the form $d(x,a) \leqslant r$. By an open ball of radius r we mean an equivalence class of the form $d(x,a) < r$.
3) Any closed ball of radius r is partitioned into infinitely many open balls of radius r.

Example 2.3.48. The theory of ultrametric spaces defined above is a complete theory T in the given language which has quantifier elimination and is separably categorical. Fix a saturated model $M \models T$. Let $B = \{x \in M : d(x,c) \leqslant r\}$ be the closed ball in M centered at c of radius r, where $0 < r < 1$. Then B is not a definable set.

Proof. We just discuss the last statement. Assume to the contrary that $dist(x, B)$ is a definable predicate. Informally, the idea will be to show that there is a type p such that $dist(p, B) = 0$ but $p = \lim_i p_i$ in the logic topology such that for all i, the quantities $dist(p_i, B)$ are bounded away from 0.

By the assumption there is a continuous function $F : S_1(T) \to [0,1]$ such that $F(p) = dist(a, B)$ for any a realizing p. Fix r. Let $r_i > r$ for $i = 1, 2, \ldots$ and $r = \inf\{r_i | i = 1, 2, \ldots\}$. Let $a_i \in M$ be such that $d(a_i, B) = r_i$. Let $\mathcal{U}$ be an ultrafilter on ω and $a = \lim_{i \to \mathcal{U}} a_i$ in M. Namely for any formula $\varphi(x)$, $\varphi(M)(a) = \lim_{i \to \mathcal{U}} \varphi(M)(a_i)$. Note that as M is saturated we are just positing that a type is realized. Let $p = \text{tp}(a)$. Hence $d(c, a) = r$. So $a \in B$, $dist(a, B) = 0$ and $F(\text{tp}(a)) = 0$.

If F were continuous there would be an open neighborhood O of p in $S_1(T)$ such that $\forall q \in O$, $F(q) \in [0, r')$ for some $0 < r' < r$. However every such open neighborhood O must contain some $p_i = \text{tp}(a_i)$ and we must have $d(a_i, B) = F(p_i) \geqslant r$. This is because for all $e \in B$, $d(e, c) \leqslant r \leqslant r_i$ and $d(c, a_i) = r_i$. By (*), $d(c, a_i) \leqslant max\{d(c, e), d(e, a_i)\}$ i.e. $r_i \leqslant max\{r, d(e, a_i)\}$ so $d(e, a_i) \geqslant r_i$. So for all i, $d(a_i, B) \geqslant r$ and so we have derived the desired contradiction. □

This also gives an example where the "logic" topology on the type space is strictly coarser than the d-metric topology.

2.4 Stability in Continuous Logic

Stability in FOL is traditionally related to categoricity (of a complete FOL theory). Recall that if T is an FOL theory in a countable language L and κ is an infinite cardinal, then T is κ-categorical if T has exactly one model of cardinality κ up to isomorphism. $\aleph_0$-categoricity is a very special case and is connected to the theory of permutation groups in the form of "oligomorphic groups" and descriptive set theory ($L_{\omega_1 \omega}$). But κ-categoricity for uncountable κ is a more fundamental notion. Morley's theorem says T is κ-categorical for some $\kappa > \aleph_0$ if and only if T is κ-categorical for all $\kappa > \aleph_0$. An important characterization is that T is κ-categorical for uncountable κ if and only if T is ω-stable and unidimensional. A theory T is called ω-stable if for all countable $M \models T$ $S_1(M)$ (complete 1-types of $\mathrm{Th}(M, a)_{a \in M}$) is countable. An important step in proving Morley's theorem is to prove that categoricity in some uncountable cardinal κ implies ω-stability.

Recall that if T is a complete FOL L-theory and $\varphi(\bar{x}, \bar{y})$ is an L-formula then $\varphi(\bar{x}, \bar{y})$ is stable (for T) if there do not exist $M \models T$ and $\bar{a}_i, \bar{b}_i \in M^n$ for $i < \omega$ such that $M \models \varphi(\bar{a}_i, \bar{b}_j)$ if and only if $i \leqslant j$ $\forall i, j$. The theory T is said to be stable if every $\varphi(\bar{x}, \bar{y})$ is stable for T. For countable theories ω-stable implies stable.

One can ask about the analogies or generalizations in continuous logic. We start with a basic example.

2.4.1 *Hilbert spaces*

A real Hilbert space is a vector space H over $\mathbb{R}$ with a real valued inner product $\langle\ ,\ \rangle$ such that $\forall r, s \in \mathbb{R}$, $\forall x, y \in H$,

1) (Linearity in the first coordinate) $\langle rx + sy, z\rangle = r\langle x, z\rangle + s\langle y, z\rangle$
2) (Symmetry) $\langle x, y\rangle = \langle y, x\rangle$
3) $\langle x, x\rangle \geqslant 0$, and $\langle x, x\rangle = 0$ iff $x = 0$
4) $||x|| := \sqrt{\langle x, x\rangle}$ is a norm which induces a metric $d(x, y) := ||x - y||$ and H is complete with respect to this metric.

If the requirement of completeness for the induced metric is dropped from condition 4), the resulting structure is called a pre-Hilbert space (whose completion will be a Hilbert space). Note that condition 2) together with condition 1) implies linearity in both coordinates of the inner product. A complex Hilbert space satisfies the same axioms except $\langle x, y\rangle \in \mathbb{C}$ and $\langle x, y\rangle = \overline{\langle y, x\rangle}$ (complex conjugate). So the norm/metric is still real valued.

There is a natural notion of isomorphism of (real) Hilbert spaces. The finite dimensional ones are (up to isomorphism) of the form $\mathbb{R}^n$ with inner product given by $\langle(x_1, ..., x_n), (y_1, ..., y_n)\rangle = x_1y_1 + x_2y_2 + ... + x_ny_n$. What about the infinite dimensional situation?

Definition 2.4.1. An orthonormal basis of a Hilbert space is a linearly independent set $\{e_i : i \in I\} \subseteq H$ such that

1. each e_i has inner product 1 with itself, i.e. $||e_i|| = 1$, and $\langle e_i, e_j\rangle = 0$ for $i \neq j$.
2. The $\mathbb{R}$-linear span of $\{e_i : i \in I\}$ is dense in H.

Fact 2.4.2. *There exists such an orthonormal basis for any H, and any two such bases have the same cardinality.*

When H is separable (has a countable dense subset) and infinite dimensional, we have another description of H as the space l^2; the space of square summable sequences

$$\{(x_1, x_2, ...) | x_i \in \mathbb{R}, |x_1|^2 + |x_2|^2 + ... \text{ converges}\}.$$

The inner product is $\langle x, y \rangle = \sum x_i y_i$ and the norm is given by $||\bar{x}|| = \sqrt{\langle x, x \rangle} = \sqrt{|x_1|^2 + |x_2|^2 + ...}$.

Given a measure space (M, μ) the space $L^2(M, \mu) = \{f : M \to \mathbb{R} \mid \int |f|^2 d\mu < \infty\}$ with inner product $\langle f, g \rangle = \int fg d\mu$ is also a Hilbert space.

Note that if H has a countable orthonormal basis $\bar{e} = (e_1, e_2, ...)$ then the $\mathbb{Q}$-linear span of $\bar{e}$ is countable. As a dense subset of a dense subset of a space is itself dense, H is separable. Likewise we see that for every cardinal $\kappa \geqslant \omega$ there exists a unique Hilbert space with density character κ, i.e. the Hilbert space with an orthonormal basis of cardinality κ.

Now we turn to the continuous logic of these spaces and describe how to view them as many sorted structures. First, note that Hilbert spaces are unbounded as metric spaces so it is convenient to consider a Hilbert space as a many sorted continuous structure with sorts: $B_n(H) = \{x : ||x|| \leqslant n\}$ indexed by $n < \omega$. We have a constant symbol for the 0 vector in B_1. We have function symbols λ_r for $r \in \mathbb{R}$ for scalar multiplication by r from $B_n(H)$ to $B_{nk}(H)$ for each n where k is the unique natural number with $k - 1 \leqslant r < k$. Also $+, -$ from $B_n(H) \times B_n(H)$ to $B_{2n}(H)$, for each n. And $\langle , \rangle : B_n(H) \times B_n(H) \to [-n^2, n^2]$, as well as $d(x, y) = ||x - y||$ on each sort where $||x|| = \sqrt{\langle x, x \rangle}$. These symbols are suitably sorted. So here the real valued relations (including the metric) are not just $[0, 1]$ valued.

On the face of it the language is uncountable, but we could restrict to λ_r for r rational.

Let us record some basic model-theoretic facts. First we can write sentences expressing that H is an infinite-dimensional Hilbert space, to get a continuous logic theory T.

Fact 2.4.3.

(i) T is κ-categorical for all $\kappa \geqslant \aleph_0$ (again, this is with respect to density character).
(ii) T is complete with QE (completeness can be seen by Vaught's Test).
(iii) T is ω-stable.

What (iii) means is that for any separable model H, $S_1(H)$ with the d-metric topology has a dense countable subset.

A corollary of (iii), which we will come back to later is:

Fact 2.4.4. *Let H be a Hilbert space and $a_i, b_i \in B_1(H)$ (although this works for any $B_n(H)$). Suppose that $\lim_{i\to\infty} \lim_{j\to\infty} \langle a_i, b_j \rangle = \alpha < \infty$ then $\lim_{j\to\infty} \lim_{i\to\infty} \langle a_i, b_j \rangle = \alpha$. This says that the formula $\langle x, y \rangle$ is stable (see Section 2.4.3).*

Rémi Jaoui pointed out how Fact 2.4.4 has many consequences and connections, including the stability of certain formulas in simple theories, lying behind the Independence Theorem and a certain account (Pillay–Starchenko) of Tao's "algebraic regularity" theorem. See Example 2.4.23 below.

As a historical note, stability was introduced into functional analysis by Krivine and Maurey [KM81]. They considered separable Banach spaces E and called E stable if the formula $||x + y||$ is stable in E. Their actual definition was: for any ultrafilters $\mathcal{U}$ and $\mathcal{V}$ on $\mathbb{N}$ and bounded sequences $a_n, b_n \in E$ for $n \in \mathbb{N}$,

$$\lim_{n\to\mathcal{V}} \lim_{m\to\mathcal{U}} ||a_m + b_n|| = \lim_{m\to\mathcal{U}} \lim_{n\to\mathcal{V}} ||a_m + b_n||.$$

This is actually "local stability in a model" (in continuous logic) which will be discussed in Section 2.4.3. Krivine and Maurey then proved that any infinite-dimensional "stable" Banach space contains ℓ^p for some $1 \leqslant p < \infty$, generalizing and recovering a theorem of Aldous. Here ℓ^p is the set of infinite sequences $(a_1, a_2, ...)$ of reals such that $\sum_i |a_i|^p$ converges, equipped with the norm $||(a_i)_i|| = (\sum_i |a_i|^p)^{1/p}$. (Let's remark in passing that ℓ^2 is a Hilbert space where the inner product of $(a_i)_i$ and $(b_i)_i$ is $\sum_n a_n b_n$.)

2.4.2 *Local type spaces*

Fix a complete CL theory T. We will work in a saturated model $\overline{M}$. M, N will denote small elementary substructures of $\overline{M}$; and A, B will denote small subsets of $\overline{M}$. So far we have only been concerned with L-formulas without parameters and type spaces $S_n(T)$. But we could also work with parameters, e.g. $S_{\overline{x}}(A)$ is the space of complete types over A in variables $\overline{x}$, so $p(\overline{x}) \in S_{\overline{x}}(A)$ means $p(\overline{x})$ is a set of formulas $\varphi(\overline{x}) \in L_A$ with parameters from A. The special case $A = M \prec \overline{M}$ is important.

Fix an L-formula $\varphi(\overline{x}, \overline{y})$.

Definition 2.4.5. By a *complete φ-type over M* we mean $\mathrm{tp}_\varphi(\overline{a}/M)$ for some $\overline{a} \in \overline{M}^n$, where $\mathrm{tp}_\varphi(\overline{a}/M)$ is the data comprising of the value of $\varphi(\overline{a}, \overline{b})$ for all $\overline{b} \in M$. Equivalently, $\mathrm{tp}_\varphi(\overline{a}/M)$ is the collection of conditions $\varphi(\overline{x}, \overline{b}) = r$ true of $\overline{a}$, as $\overline{b}$ varies in M.

We write $S_\varphi(M)$ for the set of complete φ-types over M. Moreover, for $p(\overline{x}) \in S_\varphi(M)$ and $\overline{b} \in M$, we may write $\varphi(\overline{x}, \overline{b})(p)$ or $\varphi(p, \overline{b})$ to mean $\varphi(\overline{a}, \overline{b})$ where $\overline{a}$ realizes p.

What are the topology and metric of $S_\varphi(M)$? Note that we have a forgetful map $f : S_{\overline{x}}(M) \to S_\varphi(M) : \mathrm{tp}(\overline{a}/M) \mapsto \mathrm{tp}_\varphi(\overline{a}/M)$. So we could give $S_\varphi(M)$ the quotient topology: $X \subseteq S_\varphi(M)$ is closed if and only if $f^{-1}(X)$ is closed in $S_{\overline{x}}(M)$.

What about the metric? Let $p, q \in S_\varphi(M)$ and define

$$d(p,q) = \sup_{\overline{b} \in M} \left|\varphi(\overline{x}, \overline{b})(p) - \varphi(\overline{x}, \overline{b})(q)\right|.$$

Remark 2.4.6.

1) Note that even if the basic metric in T is discrete, we may still have non-discreteness of the metric on $S_\varphi(M)$.
2) As before, the metric topology refines the "logic" topology.

Definition 2.4.7. By a *definable φ-predicate over M*, we mean a continuous function $S_\varphi(M) \to [0, 1]$.

(Note that a continuous function $\psi : S_\varphi(M) \to [0, 1]$ induces a continuous function $\psi \circ f : S_{\overline{x}}(M) \to [0, 1]$, where f is as above.)

There is no harm in calling such a definable φ-predicate over M, a φ-formula over M. It is a fact that such φ-formulas are uniformly approximated by formulas of the form $c(\varphi(\overline{x}, \overline{b}_0), \ldots, \varphi(\overline{x}, \overline{b}_{n-1}))$, $\overline{b}_0, \ldots, \overline{b}_{n-1} \in M$ and $c : [0, 1]^n \to [0, 1]$ continuous. And even better, a φ-formula is of the form $c(\varphi(\overline{x}, \overline{b}_i) : i = 0, 1, 2, \ldots)$ where $c : [0, 1]^\omega \to [0, 1]$ is continuous.

Let us remark that there is a notion of a complete φ-type over a set A, which is more complicated. In the FOL case, a φ-formula over A is a formula $\psi(\overline{x}) \in L_A$ which is equivalent to a Boolean combination of $\varphi(\overline{x}, \overline{b}_i)$'s, $\overline{b}_i \in \overline{M}$.

Now here are some very basic notions from FOL adapted to continuous logic. Fix $\varphi(\overline{x}, \overline{y}) \in L$.

Definition 2.4.8.

1) Let $p(\overline{x}) \in S_\varphi(M)$. We say that p is a *definable* if the map taking $\overline{b} \in M$ to $\varphi(\overline{x}, \overline{b})(p)$ is induced by a definable predicate $\psi(\overline{y})$ over M, i.e. $\psi : S_{\overline{y}}(M) \to [0,1]$ continuous and for $\overline{b} \in M$, $\varphi(\overline{x}, \overline{b})(p) = \psi(\text{tp}(\overline{b}/M))$.
2) Let $M \prec N$, $p(\overline{x}) \in S_\varphi(N)$. We say that $p(\overline{x})$ is *finitely satisfiable in* M, if whenever $U_0, \ldots, U_{n-1}$ are open subsets of $[0,1]$, $\overline{b}_0, \ldots, \overline{b}_{n-1} \in N$ and $\varphi(\overline{x}, \overline{b}_i)(p) \in U_i$ for $i = 0, \ldots, n-1$, then there is $\overline{a} \in M^n$ such that $\varphi(M)(\overline{a}, \overline{b}_i) \in U_i$, $i = 0, \ldots, n-1$.

Remark 2.4.9.

1) $p(\overline{x}) \in S_{\overline{x}}(M)$ is definable if, for all $\varphi(\overline{x}, \overline{y})$, $p \upharpoonright \varphi$ in $S_\varphi(M)$ is definable.
2) Note that, if $p(\overline{x}) \in S_\varphi(M)$ is definable, then the relevant definable predicate $\psi(\overline{y})$ over M is unique.
   Proof: Let $\psi'(y) : S_{\overline{y}}(M) \to [0,1]$ be another witness for definability of p, so for $\overline{b} \in M$, $\psi(\overline{b}) = \psi'(\overline{b}) = \varphi(\overline{x}, \overline{b})(p)$. But $\{\text{tp}(\overline{b}/M) : \overline{b} \in M\}$ is dense in $S_{\overline{y}}(M)$, so by continuity $\psi = \psi'$.
3) Also, if $p(\overline{x}) \in S_\varphi(M)$ is definable, witnessed by $\psi(\overline{y})$, then we can apply $\psi(\overline{y})$ to $M' \succ M$ to obtain $p'(\overline{x}) \in S_\varphi(M')$ such that $p \subseteq p'$.

Everything in this section extends to the case where φ is replaced by finite collection Δ of formulas and we have the type spaces $S_\Delta(M)$.

Some things not treated in our exposition are imaginaries and "bounded closure" in CL. These are analogous to hyperimaginaries and $bdd^{heq}(A)$ in FOL. (Unfortunately in the CL literature there is a somewhat formal adaptation of FOL notions to continuous logic, and the expression "algebraic closure" is used for what should be called "bounded closure".)

2.4.3 *Local stability*

In what follows T is a complete CL theory in language L. Fix an L-formula $\varphi(\overline{x}, \overline{y})$.

Definition 2.4.10.

1) Let $M \models T$. We say $\varphi(\overline{x}, \overline{y})$ is *ϵ-stable in* M if there do not exist $\overline{a}_i$, $\overline{b}_i$ in M, $i < \omega$, such that
$$|\varphi(M)(\overline{a}_i, \overline{b}_j) - \varphi(M)(\overline{a}_j, \overline{b}_i)| \geqslant \epsilon \text{ for all } i < j.$$
2) $\varphi(\overline{x}, \overline{y})$ is stable in M if it is ϵ-stable in M for all ϵ.
3) $\varphi(\overline{x}, \overline{y})$ is *stable for* T if $\varphi(\overline{x}, \overline{y})$ is stable in M for all $M \models T$ (equivalently, stable in M for some saturated model $M \models T$).

Note. Suppose $\varphi(\overline{x},\overline{y})$ is an FOL formula (i.e. $\{0,1\}$-valued). Then the stability of φ in M means the following: there do not exist $\overline{a}_i,\overline{b}_i$ in M, $i<\omega$, such that $M\models\varphi(\overline{a}_i,\overline{b}_j)$ if and only if $i\leqslant j$; and there do exist $\overline{a}_i,\overline{b}_j$ in M, $i<\omega$, such that $M\models\neg\varphi(\overline{a}_i,\overline{b}_j)$ if and only if $i\leqslant j$. Then (using compactness) $\varphi(\overline{x},\overline{y})$ is stable for T if there do not exist $\overline{a}_i,\overline{b}_i$, $i<\omega$ in saturated M, such that, $M\models\varphi(\overline{a}_i,\overline{b}_j)$ if and only if $i\leqslant j$.

Lemma 2.4.11. *Fix $M\models T$ and $\varphi(\overline{x},\overline{y})$. The following are equivalent.*

1. *$\varphi(\overline{x},\overline{y})$ is stable in M.*
2. *Let $\overline{a}_i,\overline{b}_i\in M$, $i<\omega$. Suppose $\lim_j\lim_i\varphi(M)(\overline{a}_i,\overline{b}_j)$ and $\lim_j\lim_i\varphi(\overline{a}_i,\overline{b}_j)$ both exist. Then they are equal.*
3. *Let $\overline{a}_i,\overline{b}_i\in M$ for $i<\omega$. Let $\mathcal{U},\mathcal{V}$ be ultrafilters on ω. Then*

$$\lim_{i\to\mathcal{U}}\lim_{j\to\mathcal{V}}\varphi(M)(\overline{a}_i,\overline{b}_j)=\lim_{j\to\mathcal{V}}\lim_{i\to\mathcal{U}}\varphi(M)(\overline{a}_i,\overline{b}_j).$$

Proof. Routine manipulations. □

Lemma 2.4.12. *The following are equivalent:*

1. *$\varphi(\overline{x},\overline{y})$ is stable for T.*
2. *Whenever $((\overline{a}_i,\overline{b}_i):i<\omega)$ is an indiscernible sequence in any model, then for $i\leqslant j$, $\varphi(M)(\overline{a}_i,\overline{b}_j)=\varphi(M)(\overline{a}_j,\overline{b}_i)$.*

(Recall $(\overline{a}_i:i<\omega)$ is *indiscernible in M* if, for all $i_0<\cdots<i_{n-1}$, $j_0<\cdots<j_{n-1}$, $\mathrm{tp}_M(\overline{a}_{i_0},\ldots,a_{i_{n-1}})=\mathrm{tp}_M(\overline{a}_{j_0},\ldots,a_{j_{n-1}})$. If $((\overline{a}_i,\overline{b}_i):i<\omega)$ is indiscernible, then $\varphi(\overline{a}_i,\overline{b}_j)=\varphi(\overline{a}_{i'},\overline{b}_{j'})$ for all $i<j$, $i'<j'$.)

Lemma 2.4.13. *Let $\varphi_0(\overline{x},\overline{y}_0),\ldots,\varphi_{n-1}(\overline{x},\overline{y}_{n-1})$ be stable for T. Let $f:[0,1]^n\to[0,1]$ be continuous. Then $\psi(\overline{x},\overline{y}_0,\ldots,\overline{y}_{n-1})=f(\varphi_0(\overline{x},\overline{y}_0),\ldots,\varphi_{n-1}(\overline{x},\overline{y}_{n-1}))$ is stable.*

Also, if $\varphi_i(\overline{x},\overline{y}_i)$ is stable for T for $i<\omega$, and $f:[0,1]^\omega\to[0,1]$ is continuous, then $f((\varphi(\overline{x},\overline{y}_i):i<\omega))$ is stable in $(x,\overline{y}_0,\ldots)$.

We will now deduce a fundamental result from a theorem of Grothendieck, although we could also give a direct proof of it similar to Proposition 1.3.7 in the Stability chapter. We first state a special case of Theorem 6 in Grothendieck's "Critères de compacité dans les espaces fonctionneles generaux" [Gro52]. (The whole connection of Grothendieck and stability was first noticed by Ben Yaacov [Ben14], although the full story, stability in a model, was not told there.)

Theorem 2.4.14 (Grothendieck). *Let X be a compact Hausdorff space and $X_0 \subseteq X$ a dense subset. Let $A \subseteq C(X, [0,1])$ such that, if $f_i \in A$, $x_i \in X_0$ for $i < \omega$, then* $\lim_i \lim_j f_i(x_j) = \lim_j \lim_i f_i(x_j)$ *whenever all limits exists.* Then *if f is in* $\mathrm{cl}(A)$ *in* $[0,1]^X$ *with the pointwise (or Tychonoff) topology, then f is continuous.*

We will be discussing φ-types when φ is stable. It is normal to include the metric d in the picture (in the same way that in FOL we include equality). What this means is that we let $\Delta = \{\varphi(\bar{x}, \bar{y}), d(\bar{x}, \bar{z})\}$ where $d(\bar{x}, \bar{z})$ is defined to be the max of the $d(x_i, z_i)$. So by $S_\varphi(M)$ we really mean $S_\Delta(M)$. We let the reader make appropriate adjustments.

Corollary 2.4.15. *Let $M \models T$ and suppose $\varphi(\overline{x}, \overline{y})$ is stable in M. Let $N \succ M$ be saturated, and let $p(\overline{x}) \in S_\varphi(N)$ be finitely satisfiable in M. Then $p(\overline{x})$ is definable over M. Moreover, it is definable by a definable φ^*-predicate over M, where $\varphi^*(\overline{y}, \overline{x}) = \varphi(\overline{x}, \overline{y})$.*

Proof. This is just a translation, but nevertheless interesting. As said above $\varphi^*(\overline{y}, \overline{x}) = \varphi(\overline{x}, \overline{y})$, where now $\bar{y}$ is the "variable variable". So a φ^*-formula over M is given by a continuous function $S_{\varphi^*}(M) \to [0,1]$. Let $X = S_{\varphi^*}(M)$, and let $X_0 = \{\mathrm{tp}_{\varphi^*}(\overline{b}/M) : \overline{b} \in M\}$, which is dense in X. For $\overline{a} \in M$ and $q(\overline{y}) \in X$, define $f_{\overline{a}} : X \to [0,1]$ by $f_{\overline{a}}(q) = \varphi(\overline{a}, \overline{y})(q) = \varphi(\overline{a}, \overline{b})$ for some/any realization $\overline{b}$ of q in $\overline{M} \succ M$. If $q = \mathrm{tp}_{\varphi^*}(\overline{b}/M)$, $\overline{b} \in M$, this is precisely $\varphi(M)(\overline{a}, \overline{b})$. So take $A = \{f_{\overline{a}} : X \to [0,1]\}, \overline{a} \in M\}$. Then the double limit condition in Theorem 2.4.14 says precisely "$\varphi(\overline{x}, \overline{y})$ is stable in M".

Now let $p(\overline{x}) \in S_\varphi(N)$ be finitely satisfiable in M. We want to view $p(\overline{x})$ as a map $f_p : X \to [0,1]$ which is moreover in the closure of A. Why and how? For $\overline{b} \in N$, $\varphi(\overline{x}, \overline{b})(p)$ depends only on $\mathrm{tp}_{\varphi^*}(\overline{b}/M)$. Because, suppose $\overline{b}' \in N$ with $\mathrm{tp}_{\varphi^*}(\overline{b}'/M) = \mathrm{tp}_{\varphi^*}(\overline{b}/M)$, and assume for a contradiction that $\varphi(\overline{x}, \overline{b})(p) = r < s = \varphi(\overline{x}, \overline{b}')(p)$. Choose $r \in I$, $s \in J$, with $I, J \subseteq [0,1]$ open disjoint. So $\varphi(\overline{x}, \overline{b}) \in I$, $\varphi(\overline{x}, \overline{b}') \in J$ are "open conditions" in p. By finite satisfiability of p in M, there is $\overline{a} \in M$ such that $\varphi(\overline{a}, \overline{b}) \in I$ and $\varphi(\overline{a}, \overline{b}') \in J$, contradicting $\mathrm{tp}_{\varphi^*}(\overline{b}/M) = \mathrm{tp}_{\varphi^*}(\overline{b}'/M)$.

Hence we can consider $p \in S_\varphi(N)$ as a function $f_p : X \to [0,1]$ where $f_p(q) = \varphi(x, \overline{b})(p)$ for some/any $\overline{b}$ in N realizing q.

Claim. $f_p \in cl(\{f_{\overline{a}} : \overline{a} \in M\})$.

A typical open neighborhood U of f_p in $[0,1]^X$ is given by $q_0, \ldots, q_{n-1} \in X$ and open intervals $I_0, \ldots, I_{n-1} \subseteq [0,1]$ such that $U = \{f : f(q_0) \in I_0, \ldots, f(q_{n-1}) \in I_{n-1}\}$ and $f_p \in U$. Let $\overline{b}_0, \ldots, \overline{b}_{n-1}$ realize $q_0, \ldots, q_{n-1}$ respectively in N. Then the "open conditions" $\varphi(\overline{x}, \overline{b}_0) \in I_0, \ldots, \varphi(\overline{x}, \overline{b}_{n-1}) \in I_{n-1}$ are in $p(\overline{x})$, so by finite satisfiability there exists $\overline{a} \in M$ such that $\varphi(\overline{a}, \overline{b}_0) \in I_0, \ldots, \varphi(\overline{a}, \overline{b}_{n-1}) \in I_{n-1}$, which means that $f_{\overline{a}}(q_0) \in I_0, \ldots, f_{\overline{a}}(q_{n-1}) \in I_n$, i.e. $f_{\overline{a}} \in U$. This proves the Claim.

So we can apply Theorem 2.4.14 to conclude that $f_p : X \to [0,1]$ is continuous. As $X = S_{\varphi^*}(M)$ this means precisely that f_p is given by a φ^*-formula $\psi(\overline{y})$ over M. So for $\overline{b} \in N$, $\varphi(\overline{x}, \overline{b})(p) = f_p(\text{tp}_{\varphi^*}(\overline{b}/M)) = \psi(\overline{b})$. □

Remark 2.4.16. In Theorem 13.4 in [BU10], they point out that the φ^*-formula $\psi(\overline{y})$ is a "positive combination" of instances of $\varphi^*(\overline{y}, \overline{x})$, i.e. of the form $f((\varphi(\overline{a}_i, \overline{y}) : i < \omega))$, where $f : [0,1]^\omega \to [0,1]$ is continuous and increasing in each coordinate.

Corollary 2.4.17. *Let $\varphi(\overline{x}, \overline{y})$ be stable for T. Then*

1. *For any $M \models T$ and $p(\overline{x}) \in S_\varphi(M)$, p is definable by a φ^*-formula ψ_p over M. Moreover, if $N > M$, then applying ψ_p to N gives $p' \in S_\varphi(N)$ extending $p(\overline{x})$ which will also be the unique extension of p to a complete φ-type over N which is finitely satisfiable in M.*
2. *(Local forking symmetry) Let $M \models T$, $p(\overline{x}) \in S_\varphi(M)$, $q(\overline{y}) \in S_{\varphi^*}(M)$. Note $\varphi^*(\overline{y}, \overline{x})$ is also stable, so let $\chi_q(\overline{x})$ be the φ^*-definition of q, a φ-formula over M. Then $\chi_q(\overline{x})(p) = \psi_p(\overline{y})(q)$.*

Proof. 1. is obvious. 2. is a bit subtle and uses the full strength of Corollary 2.4.15. □

Remember $\|M\|$ is the density character of M, $\|S_\varphi(M)\|$ is the minimum cardinality of a dense subset with respect to the d-metric. The following is Proposition 7.7 of [BU10]. However the latter seems to be incomplete; in their proof of 2 implies 3, stability is assumed, at least notationwise.

Proposition 2.4.18. *Fix a complete theory T in language L, and $\varphi(\overline{x}, \overline{y})$ an L-formula. The following are equivalent:*

1. *$\varphi(\overline{x}, \overline{y})$ is stable for T.*
2. *For all $M \models T$ and $p(\overline{x}) \in S_\varphi(M)$, $p(\overline{x})$ is definable.*
3. *For all $M \models T$ such that $\|M\| \geqslant |L|$, $\|S_\varphi(M)\| \leqslant \|M\|$.*

4. *There is $\lambda \geqslant |L|$ such that, for all $M \models T$ such that $\|M\| \leqslant \lambda$, $\|S_\varphi(M)\| \leqslant \lambda$.*

Proof. 1. $\Rightarrow$ 2. is Corollary 2.4.17.

For 2. $\Rightarrow$ 3. we give the sketch of a proof. So, let $\lambda = \|M\|$. Now, for every $p(\overline{x}) \in S_\varphi(M)$ there is $\psi_p(\overline{y})$ a definable predicate over M defining p. The assumption $\|M\| \geqslant \|L\|$ implies that among the ψ_p there is a dense subset $\{\psi_{p_i} : i \in I\}$ of cardinality $\leqslant \lambda$. But then $\{p_i(\overline{x}) : i \in I\}$ is dense in $S_\varphi(M)$. (A few things have to be checked here.)

3. $\Rightarrow$ 4. is clear.

For 4. $\Rightarrow$ 1. This is standard adaptation of the FOL case. Assume $\varphi(\overline{x}, \overline{y})$ is unstable and fix $\lambda \geqslant |L|$. Let μ be the least cardinal such that $2^\mu > \lambda$ (so $\mu \leqslant \lambda$). $2^{<\mu}$ is the set of sequences of 0's and 1's of length $< \mu$, so $|2^{<\mu}| \leqslant \lambda$. Consider $I = 2^{<\mu}$ with the lexicographical order (a total ordering). By Theorem 2.4.12 in the monster model $\overline{M}$ saturated model we can find $(\overline{a}_i, \overline{b}_i)_{i \in I}$ indiscernible, and $r < s$ such that $i < j$ implies $\varphi(\overline{a}_i, \overline{b}_j) = r$ and $i > j$ implies $\varphi(\overline{a}_i, \overline{b}_j) = s$. Let M be an elementary substructure of $\overline{M}$ containing $\{\overline{b}_i : i \in I\} = \{\overline{b}_\eta : \eta \in 2^{<\mu}\}$ with $\|M\| \leqslant \lambda$ (by downward Lowenheim–Skolem). By compactness we can find in $\overline{M}$, for $\eta \in 2^\mu$, $\overline{a}_\eta$ such that $\varphi(\overline{a}_\eta, \overline{b}_i) = r$ if $\eta < i$, and $\varphi(\overline{a}_\eta, \overline{b}_i) = s$ if $\eta > i$. Let $p_\eta(\overline{x}) = \mathrm{tp}_\varphi(\overline{a}_\eta/M) \in S_\varphi(M)$. The p_η are distinct and there are $2^\mu > \lambda$ such types. For $\eta_1 \neq \eta_2$, $d(p_{\eta_1}, p_{\eta_2}) \geqslant r - s$, hence the density character of $S_\varphi(M)$ is $> \lambda$. □

Definition 2.4.19. Given T a complete CL theory in L, T is *stable* if every L-formula $\varphi(\overline{x}, \overline{y})$ is stable for T.

Fact. *It is enough to consider formulas $\varphi(x, \overline{y})$ where x is a single variable. Also stability of T implies any T-definable predicate in free variables $\overline{x}, \overline{y}$ is stable.*

Fix $\overline{M}$ a saturated model and small $M \prec \overline{M}$. Let $\overline{c}$ be a tuple in $\overline{M}$, $B \supseteq M$ a parameter set. We say that $\overline{a}$ is *independent from B over M* if $\mathrm{tp}(\overline{a}/B)$ is definable over M, i.e. for each L-formula $\varphi(\overline{x}, \overline{y})$, there is some definable predicate $\psi(\overline{y})$ over M such that, for all $\overline{b}$ in B, $\varphi(\overline{a}, \overline{b}) = \psi(\overline{b})$.

We just summarize the situation for stable T.

Proposition 2.4.20. *Assume T is stable, $\overline{M} \models T$ saturated.*

1. *(Existence) Given $\overline{a}$ and $B \supseteq M$, there is $\overline{a}'$ such that* $\mathrm{tp}(\overline{a}'/M) = \mathrm{tp}(\overline{a}/M)$ *and $\overline{a}'$ is independent from B over M.*

2. *(Symmetry)* $\overline{a}$ *is independent from* $\overline{b}M$ *over* M *if and only if* $\overline{b}$ *is independent from* $\overline{a}M$ *over* M.
3. *(Transitivity) Given* $M \subseteq N \subseteq B$, $\overline{a}$ *is independent from* B *over* M *if and only if* $\overline{a}$ *is independent from* B *over* N *and* $\overline{a}$ *is independent from* N *over* M.
4. *(Uniqueness) Given* $\overline{a}, \overline{a}'$ *and* $B \supseteq M$ *such that* $\text{tp}(\overline{a}/M) = \text{tp}(\overline{a}'/M)$ *and each of* $\overline{a}$, $\overline{a}'$ *is independent from* B *over* M, *then* $\text{tp}(\overline{a}/B) = \text{tp}(\overline{a}'/B)$.

Remark 2.4.21.

1. Let $p = \text{tp}(\overline{a}/B)$, $p_0 = \text{tp}(\overline{a}/M)$. When $\overline{a}$ is independent from B over M, we say "p is the non-forking extension of p_0" (see part 3 of this Remark).
2. We can extend this to the situation where M is a set $A \subseteq B$, preserving 1–3 but not 4.
3. Here is the connection with "dividing": $\overline{a}$ is independent from B over M if $\text{tp}(\overline{a}/B)$ does not divide over M. Let $p(\overline{x}) = \text{tp}(\overline{a}/B)$. We say that $p(\overline{x})$ *does not divide over* M if, for all $\overline{b} \subseteq B$ and every formula $\varphi(\overline{x}, \overline{y})$ such that $\varphi(\overline{a}, \overline{b}) = r$, and M-indiscernible sequence $(\overline{b}_0, \overline{b}_1, \ldots)$, where $\overline{b}_0 = \overline{b}$, there is $\overline{a}'$ such that $\varphi(\overline{a}', \overline{b}_i) = r$ for all $i < \omega$. In the stable context, dividing and forking coincide.

Let us revisit Hilbert spaces in the light of the above results. Let T be the theory of infinite-dimensional Hilbert spaces as described earlier. We mentioned above that T is complete and categorical in all infinite cardinalities. This implies stability, on general grounds. But we will try to see stability more directly.

Fact 2.4.22. *Let* H_1 *be a Hilbert subspace of the Hilbert space* H. *Then* $H = H_1 \bigoplus H_1^{\perp}$, *as a Hilbert space, where* $H_1^{\perp} = \{y \in H : \langle x, y \rangle = 0$ *for all* $x \in H_1\}$, *the orthogonal complement of* H_1.

Let us check the stability of the formula $\varphi(x, y) := \langle x, y \rangle$ in T. Let $\overline{M}$ be a monster model, $M \prec \overline{M}$. We have to check that, for all $a \in \overline{M}$, $p = \text{tp}_{\varphi}(a/M)$ is definable (and then use Proposition 2.4.18). We may assume $a \notin M$. Let $\bar{M} = M \bigoplus M^{\perp}$ as in the fact above. Let $a = b + c$, where $b \in M$ and $c \in M^{\perp}$. We write $b = P_M(a)$ (projection of a on M). Then we see that, for any $d \in M$, $\langle a, d \rangle = \langle b, d \rangle$ (why?), and this gives our definition for $tp_{\varphi}(a/M)$.

This has many consequences for even FOL theories. Here is one.

We learned of it from Rémi Jaoui who heard it in a graduate course by Hrushovski in Paris.

Example 2.4.23. Let M be an FOL structure in language L, say saturated. Let μ_x be a Keisler measure in sort x over M, i.e. a $[0,1]$-valued finitely additive probability measure on subsets of the x sort in M that are definable with parameters in M. Let $\varphi(x,\overline{y})$, $\psi(x,\overline{z})$ be L-formulas. For $\overline{b},\overline{c}$ in M, let $R(\overline{b},\overline{c}) = \mu(\varphi(x,\overline{b}) \wedge \psi(x,\overline{c})) \in [0,1]$. Expand M to a CL-structure $M' = (M,R)$, R on $M^m \times M^n$. Then R is stable for $\mathrm{Th}(M')$.

Proof. Consider $S_x(M)$. Then μ induces a Borel probability measure on $S_x(M)$. Let $H = L^2(\mu)$, the Hilbert space of square integrable functions on $S_x(M)$, i.e. $f : S_x(M) \to \mathbb{R}$ measurable such that $\int |f|^2 d\mu < \infty$, with $\langle f,g\rangle = \int fg d\mu$.

We have just seen that $\langle,\rangle$ is stable for T (theory of Hilbert spaces). For $\overline{a} \in M^n$, let $f_{\overline{a}} : S_{\overline{x}}(M) \to \{0,1\}$ be $f_{\overline{a}}(p) = \varphi(\overline{x},\overline{a})(p)$ (= 0 or 1). Likewise $g_{\overline{b}} = \psi(\overline{x},\overline{b})(p)$. Then

$$\mu(\varphi(\overline{x},\overline{a}) \wedge \psi(\overline{x},\overline{b})) = R(\overline{a},\overline{b}) = \int f_{\overline{a}} g_{\overline{b}} d\mu = \langle f_{\overline{a}}, g_{\overline{b}}\rangle$$

for all $\overline{a} \in M^m$, $\overline{b} \in M^n$. Hence R is stable for $\mathrm{Th}(M')$. □

We also mentioned earlier (Fact 2.4.3) that T has QE. A proof can be given using the following lemma, which itself can be proved by a back and forth argument.

Lemma 2.4.24. *Let $a_0,\ldots,a_{n-1}$, $b_0,\ldots,b_{n-1} \in \overline{M} \models T$. Let $A \subseteq \overline{M}$. Let $\overline{A}$ be the norm closure of the linear span of A. Then* $\mathrm{tp}(a_0,\ldots,a_{n-1}/A) =$ *$tp(b_0,\ldots,b_{n-1}/A)$ in $\overline{M}$ if and only if $P_{\overline{A}}(a_i) = P_{\overline{A}}(b_i)$, $i < n$, and $\langle a_i,a_j\rangle = \langle b_i,b_j\rangle$ for all $i,j < n$.*

Note that we can now deduce stability of the continuous logic theory of Hilbert spaces: we have seen stability of the inner product. Stability of the metric is automatic, and that of the (graded) versions of $+,-$ and scalar multiplication is easy. Hence by QE the theory is stable.

Finally in this section we discuss some anomalies of CL-stability compared with FOL stability.

Recall that in the stable FOL context a complete type $p(x) \in S(M)$ is said to have U-rank 1 (also called "minimal") if p is not realized in M and the only forking extensions of p over larger models are realized types. There always exist U-rank 1 types in a stable FOL theory (by a union of chains argument and compactness). The notion of U-rank 1 type also makes sense in CL-stability.

Lemma 2.4.25. *In the theory T of infinite-dimensional Hilbert spaces there are no U-rank 1 types.*

Proof. Write the monster model $\bar{M}$ as $M \bigoplus M^{\perp}$ (where M is a small elementary substructure of $\bar{M}$) and let $a \notin M$. So we are considering $p = tp(a/M)$. Let $a = b+c$ be such that $b \in M$ and $c \in M^{\perp}$. So $c \neq 0$. Now we can find small M_1 such that $M^{\perp} = M_1 \bigoplus M_1^{\perp}$ such that $c = c_0 + c_1$ and $c_0, c_1 \neq 0$, and let $N = M \bigoplus M_1$ an elementary extension of M. Now note $\mathrm{tp}(a/N)$ is not definable over M. In fact, $\langle a, d\rangle = \langle b, d\rangle$ for all $d \in M$. But $P_N(a) = b + c_0$ and $\langle a, b + c_0\rangle \neq \langle b, b + c_0\rangle$. So $tp(a/N)$ is not definable over M, namely it is NOT the case that $\langle a, d\rangle = \langle b, d\rangle$ for all $d \in M_1$, as a has different projections on M and N. Specifically $\langle a, b + c_0\rangle \neq \langle b, b + c_0\rangle$. And of course $tp(a/N)$ is not a realized type. □

2.4.4 *Keisler measures and continuous stable regularity*

I finished the stability theory chapter with an account of the stable regularity lemma, using pseudofinite methods and local stability. In recent work with Chavarria and Conant we proved analogous results in a "functional framework" using also pseudofinite methods and local stability in continuous logic. I will describe some of the key points without proof, referring the reader to the paper [CCP24] for more details as well as references.

In the stability chapter I described the classical Szemeredi (graph) regularity lemma, partitioning finite (bipartitite graphs) into a small number of subgraphs, each of which is regular. Regularity meaning that sufficiently large subgraphs have the same density. (Of course these are all approximate statements relative to a small ϵ.) Szemeredi regularity has already been recast in an analytic framework by several combinatorialists. Here the finite graph (V, W, E) is replaced by a function $f : V \times W \to [0, 1]$.

What we call stable graph regularity is Szemeredi regularity where the graph relation E is assumed to be "uniformly stable" meaning k-stable for a given k. In this case the partition theorem is improved in various ways. Among these is that regularity is replaced by homogeneity, meaning the complete graph or empty (no edges) graph.

Model-theoretic (essentially nonstandard analysis) proofs of these results go through passing to a pseudofinite graph equipped with its nonstandard counting measure. In the general case, Radon–Nikodym can be applied. In the stable case it is formula-by-formula stability.

Here I want to describe the analytic version of stable regularity, where we consider $f : V \times W \to [0,1]$ but assume some kind of (uniform) stability of f. Viewing f as a $[0,1]$ valued relation it is clear that continuous logic stability is relevant. Obtaining the stable analytic regularity statement is far from routine and requires some new results about Keisler measures and stable formulas in continuous logic.

The whole "pseudofinite" apparatus is also somewhat delicate compared to the FOL case. Passing to a pseudofinite model (namely ultraproduct) of the finite $f : V \times W \to [0,1]$ should give us a CL-structure with f a stable relation. What exactly the pseudofinite counting measure is and how it is obtained is also delicate, as is the process of transferring results from the ultraproduct to the finite. Here I will not really discuss these issues, and just give the CL-results on Keisler measures and stable formulas. And then state the regularity statement.

Let us fix a CL-structure M (saturated if one wishes) in a language L. Fix an L-formula $\varphi(x,y)$ (where x,y are tuples of variables) and consider the local type space $S_\varphi(M)$. (As above we really mean $S_\Delta(M)$ for $\Delta = \{\varphi(x,y), d(x,y)\}$.) As above this comes with its d-metric (even though the underlying metric on M may be discrete).

Definition 2.4.26. By a Keisler φ-measure μ over M we mean a regular Borel probability measure μ on $S_\varphi(M)$.

In the FOL case a Keisler φ-measure over M is determined by its restriction to clopens, i.e. to φ-formulas over M, this being the original definition of Keisler measure. In the CL-case the analogue is a suitable linear functional on the real vector space of continuous functions from $S_\varphi(M)$ to $\mathbb{R}$. See [CCP24] for details.

Given a closed (nonempty) subset C of $S_\varphi(M)$, $diam(C) = sup\{d(p,q) : p,q \in C\}$. Note that $diam(C) = 0$ iff C is a singleton. In the FOL case, when $\varphi(x,y)$ is stable, any Keisler φ-measure over M is a "weighted" sum of Diracs. (Lemma 2.1 in [MP16] and Lemma 3.39 in the stability chapter.) We will generalize to the CL-case. In fact only the notion of δ-stable is needed rather than full stability ($0 < \delta < 1$). As at the beginning of Section 2.4.3 we define $\varphi(x,y)$ to be δ-stable (for $Th(M) = T$) if there do not exist a_i, b_i in any model N of T, $i < \omega$, such that $|\varphi(a_i,b_j) - \varphi(a_j,b_i)| \geqslant \delta$ for all $i < j$.

With all this notation and definitions we have:

Theorem 2.4.27. *Suppose that $\varphi(x,y)$ is δ-stable for $Th(M)$. Let μ be a*

Keisler φ-measure over M. Then there is a countable set $\{C_i : i \in I\}$ of closed pairwise disjoint subsets of $S_\varphi(M)$, such that

(i) Each C_i has diameter at most 2δ,
(ii) $\mu = \sum_{i \in I} \alpha_i \mu_i$, where μ_i is a Keisler φ-measure on $S_\varphi(M)$ concentrating on C_i and $\sum_{i \in I} \alpha_i = 1$.

Note that in the FOL case, taking $\delta < 1/2$, we recover the earlier result.

The proof of Theorem 4.27 in [CCP24] goes through a Cantor–Bendixon analysis. (See [BU10] where the case of $\varphi(x, y)$ fully stable was considered.) Keisler measures on spaces $S_\varphi(M)$ were studied in [BY08], and in the case that $\varphi(x, y)$ is fully stable, Theorem 2.4.27 is similar to Theorem 3.31 from [BY08].

Theorem 2.4.27 was one of the key ingredients in the proof of the continuous stable regularity theorem in [CCP24] which we will now state, after a few more definitions.

We are now concerned with V, W finite and $f : V \times W \to [0, 1]$.

Definition 2.4.28. $f : V \times W \to [0, 1]$ is (k, δ)-stable if there do not exist $a_1, ..., a_k \in V$, $b_1, ..., b_k \in W$ such that $|f(a_i, b_j) - f(a_j, b_i)| \geqslant \delta$ for all $i < j$.

The reason for the k is that an ultraproduct of (k, δ) stable f's will be δ-stable. Now for homogeneity.

Definition 2.4.29. $f : V \times W$ is (δ, ϵ)-homogeneous, witnessed by $r, s \in [0, 1]$, if

(i) for all but at most $\epsilon|W|$ many $b \in W$, for all but at most $\epsilon|V|$ many $a \in V$, we have $f(a, b)$ is within δ of r, and dually
(ii) for all but at most $\epsilon|V|$ many $a \in V$, for all but at most $\epsilon|W|$ many $b \in W$, $f(a, b)$ is within δ of s.

It follows that r is close to s, i.e. $|r - s| < 2(\delta + 2\epsilon)$.

The following is a simplified version of the main result Theorem 5.1 of [CCP24].

Theorem 2.4.30. *Suppose V, W are finite and $f : V \times W \to [0, 1]$ is a (k, δ)-stable function. Let $\epsilon > 0$. Then there are m, n depending on k, δ, ϵ, and partitions $V = V_1 \cup .. \cup V_m$, $W = W_1 \cup ... \cup W_n$, such that for each $1 \leqslant i \leqslant m$ and $1 \leqslant j \leqslant n$, $(V_i, W_j, f|(V_i \times W_j))$ is $(5\delta + \epsilon, \epsilon)$-homogeneous.*

References

[BBHU08] Ben Yaacov, Alexander Berenstein, Ward Henson, and Alexander Usvyatsov. *Model Theory For Metric Structures.* London Mathematical Society Lecture Note Series. Cambridge University Press, 2008.

[Ben14] Itaï Ben Yaacov. Model-theoretic stability and definability of types, after A. Grothendieck. *Bulletin of Symbolic Logic*, 20(4):491–496, 2014.

[BU10] Itaï Ben Yaacov and Alexander Usvyatsov. Continuous first order logic and local stability. *Transactions of the American Mathematical Society*, 362(10):5213–5213, Oct 2010.

[BY08] Itaï Ben Yaacov. Topometric spaces and perturbations of metric structures. *Logic and Analysis*, 1(235), 2008.

[Cas11] Enrique Casanovas. *Simple Theories and Hyperimaginaries.* Lecture Notes in Logic. Cambridge University Press, 2011.

[CCP24] Nicolas Chavarria, Gabriel Conant, and Anand Pillay. Continuous stable regularity. *Journal of the London Mathematical Society*, 109, 2024.

[CK85] Chen C. Chang and Howard Jerome Keisler. *Continuous model theory.* Princeton Univ. Press, 1985.

[CP23] Nicolas Chavarria and Anand Pillay. On pp-elimination and stability in a continuous logic environment. *Annals of Pure and Applied Logic*, 174, 2023.

[DCK71] D. Dacunha-Castelle and Jean-Louis Krivine. Applications des ultraproduits a l'etude des espaces et des algebres de banach. *Studia Math.*, 41:315–351, 1971.

[Gro52] A. Grothendieck. Criteres de compacite dans les espaces functionnels generaux. *American Journal of Mathematics*, 74(1):168–186, 1952.

[HIKO03] C. Ward Henson, José Iovino, Alexander S. Kechris, and Edward Odell. *Part One: Ultraproducts in Analysis*, pp. 1–114. London Mathematical Society Lecture Note Series. Cambridge University Press, 2003.

[HKP00] Bradd Hart, Byunghan Kim, and Anand Pillay. Coordinatisation and canonical bases in simple theories. *The Journal of Symbolic Logic*, 65(1):293–309, 2000.

[HKP22] Ehud Hrushovski, Krzysztof Krupinski, and Anand Pillay. Amenability, connected components, and definable actions. *Selecta Mathematica*, 28, 2022.

[HLCM74] C. Ward Henson and L. C. Moore. Nonstandard hulls of the classical Banach spaces. *Duke Mathematical Journal*, 41(2):277–284, 1974.

[HM74] C. Ward Henson and L. C. Moore. Subspaces of the nonstandard hull of a normed space. *Transactions of the American Mathematical Society*, 197:131–143, 1974.

[KM81] J. L. Krvine and B. Maurey. Espaces de banach stables. *Israel Journal of Mathematics*, 39(4), 1981.

[MP16] Maryanthe Malliaris and Anand Pillay. The stable regularity lemma revisited. *Proceedings of the American Mathematical Society*, 144:1761–1765, 2016.

[Pil96] Anand Pillay. *Geometric Stability Theory*. Oxford logic guides. Clarendon Press, 1996.

[Pil02] Anand Pillay. Lecture notes - model theory, 2002. `https://www3.nd.edu/~apillay/pdf/lecturenotes_modeltheory.pdf`.

[Pil03a] Anand Pillay. Lecture notes - applied stability theory, 2003. `https://www3.nd.edu/~apillay/pdf/lecturenotes.applied.pdf`.

[Pil03b] Anand Pillay. Lecture notes - stability theory, 2003. `https://www3.nd.edu/~apillay/pdf/lecturenotes.stability.pdf`.

[Pil08] Anand Pillay. *An introduction to stability theory*. Dover Publications, 2008.

[Ste76] Jacques Stern. Some applications of model theory in banach space theory. *Annals of Mathematical Logic*, 9(1):49, 1976.

[TZ12] Katrin Tent and Martin Ziegler. *A course in model theory*, volume 40. Cambridge University Press, 2012.

Index